D1783633

Energy, Climate and the Environment

Series Editors
David Elliott
The Open University
Milton Keynes, UK

Geoffrey Wood
School of Law
University of Stirling
Stirling, UK

The aim of this series is to provide texts which lay out the technical, environmental and political issues relating to proposed policies for responding to climate change. The focus is not primarily on the science of climate change, or on the technological detail, although there will be accounts of this, to aid assessment of the viability of various options. However, the main focus is the policy conflicts over which strategy to pursue. The series adopts a critical approach and attempts to identify flaws in emerging policies, propositions and assertions. In particular, it seeks to illuminate counter-intuitive assessments, conclusions and new perspectives. The intention is not simply to map the debates, but to explore their structure, their underlying assumptions and their limitations. The books in this series are incisive and authoritative sources of critical analysis and commentary, clearly indicating the divergent views that have emerged whilst also identifying the shortcomings of such views. The series does not simply provide an overview, but also offers policy prescriptions.

More information about this series at
http://www.palgrave.com/gp/series/14966

Gökçe Mete

Energy Transitions and the Future of Gas in the EU

Subsidise or Decarbonise

Gökçe Mete
International Energy Charter
Brussels, Belgium

Energy, Climate and the Environment
ISBN 978-3-030-32613-5   ISBN 978-3-030-32614-2   (eBook)
https://doi.org/10.1007/978-3-030-32614-2

© The Editor(s) (if applicable) and The Author(s), under exclusive license to Springer Nature
Switzerland AG 2020
This work is subject to copyright. All rights are solely and exclusively licensed by the Publisher, whether
the whole or part of the material is concerned, specifically the rights of translation, reprinting, reuse
of illustrations, recitation, broadcasting, reproduction on microfilms or in any other physical way, and
transmission or information storage and retrieval, electronic adaptation, computer software, or by
similar or dissimilar methodology now known or hereafter developed.
The use of general descriptive names, registered names, trademarks, service marks, etc. in this
publication does not imply, even in the absence of a specific statement, that such names are exempt
from the relevant protective laws and regulations and therefore free for general use.
The publisher, the authors and the editors are safe to assume that the advice and information in this
book are believed to be true and accurate at the date of publication. Neither the publisher nor the
authors or the editors give a warranty, expressed or implied, with respect to the material contained
herein or for any errors or omissions that may have been made. The publisher remains neutral with
regard to jurisdictional claims in published maps and institutional affiliations.

Cover illustration: imaginima/gettyimages

This Palgrave Macmillan imprint is published by the registered company Springer Nature Switzerland AG
The registered company address is: Gewerbestrasse 11, 6330 Cham, Switzerland

This work is dedicated to my father, without whom this book would not have seen the light of day.

Foreword

No-one would have thought many years ago that the Internal Energy Market plan, designed to improve Europe's competitiveness vis-à-vis its then rivals, the USA and Japan, would have transformed so completely into a comprehensive social rather than a mere economic project. But it has. Indeed, the speed with which it has done so in the decade since Lisbon and the sweeping character of the policy change still encourages a double-take response. Is this really where we are now? Well, it is.

To what extent this change has been driven by internal dynamics at the European Union, such as a treaty change, and perhaps a new generation of talent at the European Commission, or by a growing awareness of the urgency of acting to mitigate climate change and related problems, is impossible to state with any exactitude. Both have surely played a role and perhaps the weight of each is of no real consequence. The effect is clear enough. The energy sector and the European policy that applies to it is being driven in a different but probably more socially relevant direction than ever before.

Gas has played an ambivalent part in this transformation—in European terms we might call it the first energy transition. As the most recent of Europe's energy sources, its vast network of pipelines and related infrastructure rolled out across the continent from a first base in the northern Netherlands to points suitable for transmission and distribution of imports from Russia, Algeria and other sources. This heavy reliance on interrelated infrastructure made the industry an early target for EU competition law and for the first generation of specialist energy regulators keen to identify potential abuses of the natural monopoly power this infrastructure gave the industry.

Was it a fuel of the future, a transitional fuel to a lower-carbon world, as its lobbyists told us, or was it just another fossil fuel that played a smaller part in the carbon-intensive economy of the past, but a part, nevertheless? As our understanding of European geography and energy dependence grew with the new members from Central and East Europe, an appreciation of natural gas' contribution to energy security took hold among the European institutions. Issues of solidarity and energy security joined the familiar ones of sustainability and competitiveness, creating the special dilemma for energy policy that Dr. Mete has written about in her book. Finding a way out of this policy conundrum will require careful assessment of the benefits and costs of natural gas in Europe, as its energy economy is reshaped for a lower carbon intensity in the future; in other words, for the energy transition we have now embarked upon. The destination of this transition is different, and in many respects still unclear, but a road map is emerging. It is one that will include the use of natural gas in some form and to some extent.

In this next stage of the debate, we need to have clear and bold thinking about the future of natural gas in the European Union. Dr. Mete's book is therefore more than a sound and authoritative study. It is a timely policy intervention, which will have practical significance for government officials and commentators around Europe. The fact that years of research have gone into the study—which the subject matter deserves—will add to its impact on thinking about our choices not only for the future role of gas, and indeed energy policy but also for the way we need to approach the energy transition itself. The author's command of the diverse source materials she has examined, combined with her critical perspective and concise empirical case studies, ensure that her conclusions will be considered in both policy making, and academic study of the ongoing energy transition in Europe and more widely.

Professor Peter Cameron, Ph.D., FRSE, FCIArb, FEI
Director, Centre for Energy, Petroleum
and Mineral Law & Policy
University of Dundee
Barrister (England and Wales)
(Middle Temple) and Arbitrator
Dundee, UK

Acknowledgements

This book is the result of six years of research inspired by my general interest in the field of energy and law and policy and, in particular, the energy transition. This book is based partly on my Ph.D. which I undertook at the Centre for Energy, Petroleum, Mineral Law and Policy (CEPMLP) at the University of Dundee. I moved to Scotland in 2013, driven by the worldwide reputation in education and research excellence of the CEPMLP in this sector. The experience was challenging at times, especially after switching to full-time employment at the CEPMLP in 2016, but the amount of support provided by the University of Dundee and my colleagues at the CEPMLP has been invaluable. I would have not completed the research without the tenacious support of CEPMLP Director Professor Peter Cameron and my Ph.D. supervisors Dr. Sergei Vinogradov and Professor Andrey Konoplyanik. I am truly thankful to them for their dedicated guidance, patience and encouragement; in short, for their mentorship.

The book benefited from intellectual conversations with Professor Volker Roeben and Dr. Stephen Dow at the CEPMLP. I thank them deeply for their insightful comments and encouragement. I am also grateful to my colleagues Janeth Warden, Professor Kim Talus, Professor

Volker Roeben, Dr. Ana Elizabeth Bastida and Mr. Stephen Dow (CEPMLP) and Joanne Jones, William Macpherson, Sophia Antoniou and Rose Wang (the Extractives Hub project). Thank you for your understanding, support and tolerance over the years, and for driving me on with your constructive enthusiasm. I would also like to pass special gratitude to CEPMLP's Commercial Manager Hugh Gunn for his intellectual and emotional support over the years. I could never have imagined undertaking such demanding research in such a fulfilling way, in any institution other than the CEPMLP at the University of Dundee. I thank all my colleagues therefore, for their academic excellence and guidance and for all the opportunities provided which have broadened my knowledge in the energy sector.

The CEPMLP has also given me the opportunity to meet most inspiring peers at the Ph.D. program, among them, Dr. Erman Ugur Ozgur, Dr. Geoff Wood, Dr. Ana Maria Daza Clark, Dr. Abdoul Kabele Camara and Crystal Svanikier have all played a vital role in this achievement. I would like to thank them for their friendship. I would like to thank Dr. Geoff Wood deeply also for his robust editorial support, his invaluable comments, inspiration and encouragement over the last nine months, since I decided to publish my research.

This research would have not been complete without the knowledge gained during my 10-month fellowship at the Energy Charter Secretariat in 2015/6. The research benefited greatly from the data collected during this period, as well as intelligent conversations with my colleagues. I would like to thank the Secretariat for the opportunity, in particular Secretary-General Dr. Urban Rusnak, Dr. Matteo Barra and Kanat Botbaev, for their guidance. From the Energy Charter Secretariat, I also thank, my former colleagues, Anna Aslanidze, Claudia Nocente and Darya Varasteh for their friendship and support which has resulted in lifelong friendships.

In 2019, six months after I successfully defended my Ph.D. thesis, I have returned to the Energy Charter Secretariat as the Head of the Knowledge Center. Writing this book in Brussels, in the heart of the European Union, alongside a demanding full-time position, allowed me to interact with some of the most knowledgeable and influential stakeholders in the gas sector in the EU. Therefore, this book benefited from exchanges with colleagues from industry and practice: in particular I express gratitude to James Watson, Richard Tyler, Christian Schwarck, Graham Coop and Simone Tagliapietra for the intellectual debates we had and I sincerely thank Daria Nochevnik, Valeria Palmisano, Margot Loudon, Dennis Hesseling and Theirry Deschuyteneer for reviewing parts of my work and for their insightful comments.

Finally, I am indebted to my family and those who supported me on a personal level. I thank my father Bahadır, my mom Müberra and my sister Özge for their unconditional support, compassion, generosity and constant encouragement throughout writing this thesis and my life in general. I am also grateful to my boyfriend Daniel Rae and the Rae family for their kindness, support, encouragement and generosity.

Contents

Abbreviations

ACER	Agency for the Cooperation of Energy Regulators
ADB	Asian Development Bank
BAL NC	Balancing Network Code
Bcm	Billion Cubic Metres
BIT	Bilateral Investment Treaty
BTC	Baku Tbilisi Ceyhan
CAM NC	Capacity Allocation Mechanism Network Code
CCUS	Carbon Capture Utilisation and Storage
CEE	Central Eastern Europe
CEER	Council of European Energy Regulators
CEF	Connecting Europe Facility
CIS	Commonwealth of Independent States
CNPC	China National Petroleum Corporation
CO_2	Carbon Dioxide
DESFA	Greek National Transmission Company
DG ENER	EU Directorate General for Energy
EBRD	European Bank of Reconstruction and Development
EC	European Commission
ECA	Export Credit Agency
ECJ	European Court of Justice
ECOWAS	Economic Community of West African States

ECS	Energy Charter Secretariat
ECT	Energy Charter Treaty
EEZ	Exclusive Economic Zones
EFET	European Federation of Energy Traders
EIA	Environmental Impact Assessment
EIB	European Investment Bank
EnCT	Energy Community Treaty
ENTSO-E	European Network of Transmission Systems Operators for Electricity
ENTSOG	European Network of Transmission Systems Operators for Gas
ERDF	European Regional Development Fund
ESIF	European Structural and Investment Funds
EU	European Union
EURATOM	European Atomic Energy Community
FID	Final Investment Decision
FSU	Former Soviet Union
GATS	General Agreement on Trade in Services
GATT	General Agreement on Tariffs and Trade
GIE	Gas Infrastructure Europe
GTM	Gas Target Model
HGA	Host Government Agreement
IDA	International Development Association
IEA	International Energy Agency
IFC	International Finance Corporation
IGA	Intergovernmental Agreement
ISO	Independent System Operator
ITE	Iran Turkey Europe Pipeline
ITO	Independent Transmission Operator
ITRE	European Parliament's Committee on Information, Research and Energy
KRG	Kurdish Regional Government
LNG	Liquefied Natural Gas
MENA	Middle East and North Africa
MIGA	Multilateral Investment Guarantee Agency
MoU	Memorandum of Understanding
NBP	National Balancing Point
NGO	Non-Governmental Organisations
NICO	Naftiran Intertrade Company Sàrl

NIOC	National Iranian Oil Company
NRA	National Regulatory Authority
OIES	Oxford Institute for Energy Studies
OPAL	Ostsee-Pipeline-Anbindungsleitung
OPEC	Organization of the Petroleum Exporting Countries
OU	Ownership Unbundling
OUP	Oxford University Press
PCI	Project of Common Interest
PECI	Projects of Energy Community Interest
PMI	Project of Mutual Interest
REIO	Regional Economic Integration Organization
SCP	South Caucasus Pipeline
SCPX	South Caucasus Pipeline Expansion
SE4ALL	Sustainable Energy for All
SOFAZ	Sovereign Wealth Fund of Azerbaijan
SOS	Security of Supply
SPV	Special Vehicle Company
TANAP	Trans Anatolian Pipeline
TAP	Trans Adriatic Pipeline
TAPI	Turkmenistan-Afghanistan-Pakistan-India
TAR NC	Tariffs Network Code
TCP	Trans-Caspian Gas Pipeline
TEN	Trans-European Networks
TEU	Treaty on European Union

1

Introduction

1.1 What Role for Gas?

The dialogue on natural gas is becoming increasingly politicised. This is intrinsically tied-in with the plethora of roles envisaged for gas by a range of stakeholders: a transition fuel bridging the transition to a low carbon energy future; a necessity to ensure security and diversity of supply in an increasingly unstable geopolitical landscape; and as a polluting fossil fuel partly driving anthropogenic climate change. Indeed, some have heralded a potential 'Golden Age of Gas' which is expected to reverse the decline in gas demand brought on by the shift to a low carbon future. Within the European Union (EU), the focus of much of this book, debate surrounding gas has resulted in disparity and a lack of meaningful communication between the industry, the European Commission (EC) and climate change lobbies. This book provides an experiential assessment of the impact of energy transitions on the future of natural gas in the EU energy mix, acknowledging that natural gas will continue to play a significant role in the heating, power and industry sectors in the EU in the medium to long-term, and the role of renewables as future partners in the transition to a low carbon economy.

© The Author(s) 2020   **1**
G. Mete, *Energy Transitions and the Future of Gas in the EU*, Energy, Climate and the Environment, https://doi.org/10.1007/978-3-030-32614-2_1

In order to understand this, the book seeks to critically analyse current natural gas market policy and law in the EU. Essentially, the Third Energy Package was introduced in 2009 to promote market-based mechanisms, but in reality, its implementation has resulted in an increased role for the EU institutions in commercial decision-making. The case studies proposed in this book reveal that most of the gas infrastructure projects in the EU are not commercially attractive—often the projects that have strategic importance for security of supply (and diversification) get the go ahead, hence subsidies are available (the role of law and policy in this problem). This book questions whether gas infrastructure can be developed by the market without any subsidies with the concomitant risk of stranded assets and carbon lock-in taken by the private sector.

The problem, it is argued here, lies with the regulation and policy that leads to public funds being spent on gas infrastructure, lack of market mechanisms (for storage in particular) and the regulatory uncertainties that comes for instance with the Gas Directive amendment in 2019. Where there is a functioning market, the risks should be borne by the private sector. They can then assess whether to invest in liquified natural gas (LNG), storage and pipeline infrastructure, or diversify their portfolio (in renewable gases or renewable energy sources [RES]). Hydrogen, synthetic gas, biomass and Carbon Capture, Utilisation and Storage (CCUS) will require even higher subsidies to develop, but at least these projects are aligned with the EUs energy transitions agenda.

In the EU, currently, the recently adopted recast of Renewable Energy, Electricity Market Design and Energy Efficiency directives include elements relevant for hydrogen technologies and biomass. However in order to create a predictable environment for innovation and investment, clear definitions and binding EU level targets for renewable and low carbon gases along with Guarantees of Origin that would make the source of renewable and decarbonised gases transparent would be necessary. Guarantees of Origin would also be essential to demonstrate the carbon footprint and therefore contribution to the decarbonisation of each specific gas. These are expected to be addressed in 2020 under a new gas package with the new Commission taking office in Autumn 2019. This book is the first to cover these evolving issues, including a regulatory and policy framework for development and operation of hydrogen pipelines, injection of biomethane into the existing gas grid and for pipelines carrying carbon dioxide (CO_2) gases.

As it will be highlighted in the sections below, for the future low carbon and renewable gases, the marketability is not yet present at this stage. There are currently only small amounts of green hydrogen and biomass production in the world. This book will endeavour to analyse then how to finance decarbonisation of the gas sector in the long term while at the same time maintain a security of supply in the short and medium-term. It will do so based on the recent experience with financing existing natural gas infrastructures which continues to benefit heavily from public funds and incentives, and whether the risks could be shifted to the private sector to minimise the burden on the tax payer. This is timely. At the moment the EU Agency for the Cooperation of Energy Regulators (ACER) is gathering consultation responses to make proposals to the EC for future gas legislation which is expected to be drafted in 2020.

1.2 Importance of the Research

The EU gas market regulation has been going through an evolutionary transition since the early 1990s, one that is still ongoing. The internal market rules are ever changing with a new gas package planned for 2020. The future role of gas in a low carbon economy needs to be reflected in both new legislation and market design. This involves enhanced sector coupling between gas and electricity markets. The EU is not an island, it operates within an interconnected energy system which requires maintaining the relationship with existing suppliers and future partners in delivering new technologies that are aimed at achieving not only the United Nations (UN) Paris Climate Agreement Goals but also the UN Sustainable Development Goals (SDGs), in particular SDG 7 on access to affordable and sustainable clean energy for all and SDG 13 on climate change. It is also important to maintain the competitiveness of the EU in global markets and ensure that the industry can continue to deliver economic growth, but facilitate an energy transition in a way that takes into account social justice.[1]

The decarbonisation ambitions of the EU by 2050, which despite remaining unsubscribed by all Member States, aims to deliver a net zero carbon economy in principle. The recent refusal of Poland, Hungary and the Czech Republic with Germany siding with them to commit to climate neutrality by 2050, has to be read carefully. The cost and burden of decarbonisation is significant. With this in mind, an objective of this book is to critically analyse and address incompatibilities between internal gas market legislation and the financing obligations of natural gas infrastructure projects in light of rising import dependency for the EU in the short and medium term. The EUs ability to attract new volumes of gas is also tighter than ever before with the removal of destination clauses in LNG contracts, initially in the EU but increasingly also in Asian markets. The development of effective regional and global transportation systems and flows of energy requires cooperation among the producer, consumer and transit countries as well as project financiers. While the integration of renewable gases into the system means that security of supply considerations will become more internal in the long term, the external dimension of the energy sector will not disappear. And, as stated above, the lessons learned from legislative efforts to create a functioning gas market will play an important role in determining the speed and cost of the decarbonisation of the gas sector. The mechanisms for decision making concerning the selection of which projects to go ahead will be particularly relevant.

This book therefore aims to develop an understanding of and clarify the complex range of legislation involved within a single analytical framework for all readers from the public and private sectors and also from academia. It is also aimed at developing constructive debate on the issues of interaction between relevant instruments of international law such as Intergovernmental Agreements (IGAs) and the EU energy acquis.

---

[1]Gokce Mete and Raphael Heffron, The Social Dimension of EU Energy Law, in Delia Ferri and Fulvio Cortese (eds.), *The EU Social Market Economy and the Law* (Routledge Research in EU Law) (Routledge, 2018).

Furthermore, the principles of multi-stakeholder, medium to large-scale energy project development are essentially the same, whether the project is oil and gas extraction, electricity transmission lines, LNG projects, carbon capture or storage (CCS), nuclear or renewable energy development, as the investment will be often of a high-risk, long term character with a long period for return of investment. Hence, lessons learned from the EUs approach to law, policy and decision making as a host country will be relevant for all projects of a certain magnitude in terms of finance and long-term commitment.

It is currently true that there are not so many immediately available alternatives to natural gas to keep the lights on, ensure warm (and cool) homes and make sure industry continues to run. However, this book acknowledges that as we approach 2050, the requirement to sharply decrease CO_2 and other GHG emissions means that the role of gas infrastructure in the EU will change drastically. But what does such change mean? It may mean that there will be more gas in the Distribution System Operator (DSO) grid than in the Transmission System Operator (TSO) network. However, there is significant sunk investments made by the TSO and they will still need to be depreciated. It is likely that in order to accommodate more hydrogen and reverse renewable gas flows, large sums of money will need to be invested in refurbishing existing pipelines. Even before we get there, significant investment will be needed in the scalable production of these new technologies, in particular hydrogen production with CCUS. Even after 2050, it is highly likely that natural gas will not completely be out of the picture. We will most probably see a more flexible and diversified energy system, but one which will take many years to build from now on. Therefore, the final and perhaps the most important and central purpose of this book is to identify whether the EU will continue to subsidy natural gas projects or decarbonise the gas grid before 2050, and at what cost. In the next chapters, this book will look for answers to the following questions: how to maximise the potential of the gas infrastructure to reduce carbon emissions? What are the lessons learned from decision making experience in the natural gas sector? What is the direction of travel of the EU towards a climate neutral gas sector? How will green and low carbon gas technologies be supported? Or are the proposals to drive a growing share of hydrogen, biomethane, and synthetic methane to the system just an excuse to prolong fossil fuel operations for longer?

In summary, the research reveals the current state of the gas sector—and its transition. How to regulate technology and innovation to decarbonise gas is the most important debate in the natural gas sector in the EU currently. This book has elements of interdisciplinarity as it will use the findings of legal, policy, environmental and economic studies while carrying out an assessment of cases. It offers an opportunity for a reality check in terms of how dependent the EU is on natural gas (and third countries due to decreasing indigenous resources), and how serious is the EC in implementing the energy transitions agenda? Also, how can policy foster gas sector decarbonisation? A further reality check examined in this book centres around examining the profitability of natural gas businesses in the future, including a focus on the issues of stranded assets and sectoral sustainability. It also engages in an important debate on the cost of EU energy and offers new approaches to shift the current EU energy security paradigm that enables subsidies for fossil fuels.

Although this book focuses mainly on the EU approach to natural gas regulation and the future of gas in the EU, the findings and recommendations are relevant for a much wider geography. The share of natural gas is forecasted to increase globally—in particular in Asia with coal to gas switching. This book will draw important lessons for countries with emerging natural gas markets.

Precisely, this book seeks to fill the gap in existing literature. By aiming to present a critical account of current regulation and debate, it seeks to challenge a number of common assumptions, this book aims to address the need for an up-to-date account targeting readers in this area and provide guidance to enable the reader to quickly explore the key issues and related various facets.

1.3   Outline of the Book

Following this introduction, Chapter 2 sets the scene by discussing the importance of the research and familiarising readers with the evolution of the EU regulatory framework for natural gas and provides an initial analysis of the EU's natural gas markets attractiveness for the suppliers. Next, Chapter 3 provides a detailed analysis of the EU's energy mix and past, present and future trends of natural gas consumption, production and storage levels in the EU. This chapter provides a reality check on supply security and explores the prospect of decarbonisation of the gas grid with renewable and other low carbon gases, such as hydrogen and biogas at scale. Chapter 4 carries out a mapping exercise of natural gas subsidies and natural gas project finance in light of the Energy Transition in the EU. This chapter acquaints the reader with principles of project finance as it assesses the ability of investors to commission gas infrastructure projects based on the market. Chapter 4 also introduces how the new Sustainable Finance Package could impact future gas sector investment. Carbon pricing and fossil fuel subsidy reform recommendations are also carried out in this chapter.

All the chapters in this book use relevant case examples to support the arguments raised. Chapters 4 and 5 present around a dozen different project case studies, including, inter alia, cross-border pipelines (built within and outside Eurasia), natural gas interconnectors within the EU and LNG. Chapter 5 also explores the potential challenges waiting hydrogen and renewable gas infrastructure, a discussion which is developed further in Chapter 6 on the decision-making framework for natural gas projects in the EU and on the future role of gas. This final chapter, preceding the conclusion focuses on the legislative and regulatory aspects of the EU energy architecture. It offers both a historical and a forward-looking critical account of the energy *acquis*. This is where the fine details of the Third Energy Package, Network Codes, Gas Target Model, Energy Union and the 2019 Gas Directive Amendment are discussed. Chapter 6 describes the decision-making framework under EU natural gas market rules as an altering journey to an unknown destination. However, it delivers a number of recommendations on successful implementation of the sector coupling in the electricity and gas sectors to allow gas to play an important role in the EU's ambition to reach a net-zero carbon economy by 2050. The ongoing public discussions on a prospective 2020 Gas Package provides a timely opportunity to make decarbonisation of the gas sector a reality. In concluding, Chapter 7, while acknowledging that gas infrastructure will continue to be important, notes that a carbon free EU will not come cheap. Nor will any of the pioneering technologies, including CCS, hydrogen, and biogas, etc. offer a silver bullet to solve all issues. These are all long-term complex technologies, and considering that it took nearly two decades to establish an internal natural gas market (and it is still not complete nor free from problems), investment decisions must be taken now. The concluding chapter therefore recommends limited public resources currently being allocated to strategic, uncommercial natural gas projects to be redirected to provide market incentives for a decarbonised European gas grid of the future.

2

Setting the Scene

2.1   A Golden Age for Gas?

In 2011, the International Energy Agency (IEA) predicted a 'Golden Age of Gas'. Unlike many predictions in the fast moving energy landscape this prediction has happened.[1] Even today, forecasts up to 2023 still indicate a 1.6% annual growth in global natural gas demand.[2] Until 2015, the growth of demand for gas in Europe had been in gradual decline, partly due to the efforts required to meet energy efficiency targets, partly as a result of climate change policies and technological developments making renewables compatible with gas. The decrease in natural gas demand was also impacted by the economic recession and coal competition in power generation in the aftermath of the United States (US) shale gas revolution.[3] Falling demand has moreover

---

[1]IEA, WEO-2011 Special Report: Are We Entering a Golden Age? 06.06.2011, available via: https://webstore.iea.org/weo-2011-special-report-are-we-entering-a-golden-age.

[2]IEA, Market Report Series: Gas 2018, 26.06.2018, available via: https://webstore.iea.org/market-report-series-gas-2018.

[3]European Parliamentary Research Service, Unconventional Gas and Oil in North America, June 2014, available at: http://www.europarl.europa.eu/RegData/bibliotheque/briefing/2014/140815/LDM_BRI(2014)140815_REV1_EN.pdf.

© The Author(s) 2020
G. Mete, *Energy Transitions and the Future of Gas in the EU*, Energy, Climate and the Environment, https://doi.org/10.1007/978-3-030-32614-2_2

coincided with declining production in the EU and its strategy to reduce dependency on Russian gas (or any other single supplier). The EU was the first to remove destination clauses from long-term gas contracts and, as a result, most of the LNG, which was available to Europe before the Fukushima disaster, had to be diverted to Asian markets.[4] In 2017, after only a small rise in the demand for gas in Europe over the preceding three years, a trend for steep growth has emerged. This is the biggest growth rate seen since the first quarter of 2011.[5] This is expected to remain in view of the coal, oil and nuclear phase out in the EU.

At the same time, energy transitions to a low carbon economy is accelerating. The EU is committed to reduce its CO_2 and other GHG emissions by 2050 under the 2015 UN Paris Climate Agreement.[6] It is sufficient to say that decarbonisation is inevitable. The switch from fossil-fuels to low carbon technologies and fuels is not easy as the ultimate benefits such as successfully addressing climate change will be experienced in the long-term. Often governments fear the reaction of voters over rising costs and the risk of the 'lights going out'. The Yellow Vest protests in France proved a recent example of this.[7] Gas infrastructure is well-placed to be an integral part of this transition. Renewable gases such as biogas, gas from biomass, green hydrogen and synthetic methane from renewable energy are covered under the Third Energy Package, hence any changes to the legislative framework will have to take these factors into account and aim to encourage TSOs to be more innovative and make use of the gas infrastructure network already in situ.

Despite the fact that the demand for natural gas in the EU is expected to plateau as a result of demand side mechanisms such as energy efficiency, decreasing indigenous production and the phasing

---

[4]Council of European Energy Regulators (CEER) Report, How to Foster LNG Markets in Europe Liquefied Natural Gas Work Stream of Gas Working Group, 24.07.2019, available at: https://www.ceer.eu/documents/104400/-/-/57d62db2-db0a-e611-2a49-85703d1d54d6.

[5]Quarterly Report Energy on European Gas Markets Market Observatory for Energy DG Energy, Volume 11 (issue 1; first quarter of 2018).

[6]Energy roadmap 2050 {COM (2011) 885 final} of 15 December 2011.

[7]News article, Reuters, 'Yellow Vests' Put French Government on Spot Over Power Prices, 30.01.2019, available at: https://www.reuters.com/article/us-france-electricity/yellow-vests-put-french-government-on-spot-over-power-prices-idUSKCN1PO25Y.

out of oil, coal and nuclear, new sources of supplies and new import routes to be investigated and investment in infrastructure will all still be required. Currently all options for pipeline gas and LNG supplies from non-EU producers are expensive, needing capital intensive upstream investments and long-term contracts. Even with this, external producers would hesitate to commit to the European market due to the uncertainty of future demand. The necessary capital is not always readily available around the world, and the EU banks' ability to lend is particularly limited. Therefore, the majority of natural gas import infrastructure in the EU is subsidised with public money.

2.1.1 Why Natural Gas Is Important in the Short to Medium Term

In 2017, carbon emissions in the EU rose for the first time in seven years due to improved economic growth following the Global Financial Collapse which started in 2008.[8] Improvements in energy efficiency also drastically slowed down in 2017, as a result of weakening energy efficiency policies and the low price of coal caused by the US shale gas revolution. Gazprom set a new export target for 2018 of nearly 200 bcm of gas to Europe (including Turkey), which broke a record. These are strong indicators proving that gas infrastructure is going to be important for a long time, not only in the EU, but globally.

Annual domestic natural gas production in 35 European countries (including Turkey) currently covers only 30% of total consumption with production in Europe dropping by 30% between 1995 and 2012. In 2012, around 539 billion cubic meters (bcm) of gas was consumed by 35 European countries.[9] Total annual consumption in the European region represents approximately 16% of world consumption, making it

---

[8]Preliminary 2017 emissions under the EU's Emissions Trading Scheme (ETS) published by the European Commission, 01.04.2018, available via: https://ec.europa.eu/clima/policies/ets/registry_en#tab-0-1.

[9]Anouk Honoré, *The Outlook for Natural Gas Demand in Europe*, available via: http://www.oxfordenergy.org/wpcms/wp-content/uploads/2014/06/NG-87.pdf.

the second largest natural gas market after the US.[10] Gas consumption in Europe between 2013 and 2017 brought many surprises to the market. However, as discussed above, since 2017, after some decline in demand over the past 3 years, a new trend for growth emerged. In the first 9 months of 2017 there was a 7% increase, and by the end of 2018 demand reached a new record.[11] The first cause of this growth in demand is temperature (as global warming leads to colder winters and hotter summers), a change of 1 °C led an increase in demand for energy. The second factor is the policies of individual Member States. In the UK and Germany, for instance, gas competed with coal. This has changed in the UK as a result of the introduction of carbon taxes with gas now being the preferred energy source.[12]

The demand for gas used in the power sector has been in decline since 2013. However, in 2017 it began to increase with an extra demand of 27 bcm, of which 21.5 bcm was required for additional demand in the power generation sector. On the one hand, although gas consumption in the EU is not expected to return to the level previously seen in 2010,[13] it is undisputedly set to continue to provide a significant contribution to the European energy mix in the future. On the other hand, gas production is expected to further decline in Europe although it is on the increase in other regions such as in North America and Australia. This decline, together with rising concerns over the reliability of existing external suppliers, increases the magnitude of import dependency in Europe.

The next five to ten years will see an increasing shift from coal to gas, especially in China. Deployment of most new renewables are mainly intermittent (wind and solar), and battery technology has not yet achieved commercial scale. The EU internal energy policy aims to

---

[10]Anouk Honoré, *The Outlook for Natural Gas Demand in Europe*, available via: http://www.oxfordenergy.org/wpcms/wp-content/uploads/2014/06/NG-87.pdf.

[11]Quarterly Report Energy on European Gas Markets Market Observatory for Energy DG Energy, Volume 11 (issue 2; second quarter of 2018).

[12]I. A. Grant Wilson and Iain Staffell, Rapid Fuel Switching from Coal to Natural Gas Through Effective Carbon Pricing, *Nature Energy*, Volume 3 (2018).

[13]International Energy Agency, *World Energy Outlook, 2014*, p. 135.

deliver carbon reductions, increasing both energy efficiency and the share of renewables, but also development of additional LNG import terminals, renewable gas and consumer production. Currently, gas turbines are the best technology to provide back-up for intermittent solar and wind; therefore, gas and renewables are arguably future partners in the transition to a low carbon economy, bringing power and, with it, economic development to the developing world in Asia, Africa and Latin America.

Furthermore, end-use combustion of gas infrastructure produces only CO_2 and H_2O, which are non-toxic and odourless. Additionally, gas can be decarbonised via the capture of CO_2 from steam-reforming processes and the transformation of methane into hydrogen and solid carbon. Natural gas has high versatility; it can convey biogas, green gas and hydrogen. LNG and small-scale LNG can offer additional diversification tools for carbon intensive regions, including islands alongside off-grid and community owned renewables. LNG, pipeline and storage infrastructure has high CAPEX but very low OPEX which offers sustainability.

2.2 Evolution of the Regulatory Framework for Natural Gas and the Import Pipeline Network

At the time of writing in 2019, the most recent European internal gas market legislation is the gas legislation included as part of the Third Energy Package, which entered into force on 3 March 2011. This includes the Gas Directive concerning common rules for the internal market in natural gas (Gas Directive),[14] the Gas Regulation on conditions for access to the natural gas transmission networks

---

[14]Council Directive 2009/73/EC of the European Parliament and of the Council of 13 July 2009 concerning common rules for the internal market in natural gas and repealing Directive 2003/55/EC [2009] OJ L211/94 (Gas Directive).

(Regulation 715),[15] and the Regulation on the establishment of ACER.[16] The Package also introduced a new system for the establishment of binding European-wide Network Codes for cross-border and market integration issues, consistent with non-binding Framework Guidelines. Regulation 715 anticipates the Network Codes to cover twelve broad market issues.

The Third Energy Package brought radical changes to the existing natural gas market structure and set the framework within which the internal market needs to develop. It emerged following the 'Internal Market Review'[17] and the 'Sector Inquiry Report'[18] reflecting the EC's dissatisfaction with the implementation of earlier energy legislation. These reports, inter alia, acknowledged that enhanced security of the gas supply in the EU requires diversification of supply regions and routes. It stipulated that the best way to achieve this diversification is by establishing a competitive internal market as a means to promote investment in new infrastructure. In a nutshell, the aim was to achieve a real market with a secure supply[19] of natural gas. This objective seemed of great strategic importance for Europe given that gas is not only the cleanest fossil fuel, but also recognising limited and fast declining indigenous resources.[20]

Given Europe's dependency on supplies from third countries, the physical and regulatory transformation of its internal natural gas market must be assessed in relation to implications for the European energy

---

[15]Regulation (EC) No. 715/2009 of the European Parliament and of the Council of 13 July 2009 on conditions for access to the natural gas transmission networks and repealing Regulation (EC) No. 1775/2005.

[16]Council Regulation No. 713/2009 of the European Parliament and of the Council of 13 July 2009 establishing an Agency for the Cooperation of Energy Regulators [2009] OJ L211/1.

[17]Communication from the Commission to the European Parliament, the Council, the European Economic and Social Committee and the Committee of the Regions a single market for twenty-first century Europe, 20 November 2007, COM (2007) 724.

[18]Energy sector competition inquiry—final report—frequently asked questions and graphics, 10 January 2007, MEMO/07/15.

[19]European Commission, Energising Europe: A Real Market with Secure Supply, MEMO/07/361, 19.09.2007.

[20]IEA/OECD, Medium Term Gas Market Report (Market Analysis and Forecasts to 2019, 2014).

equation. This needs to cover all of the territories interlinked by immobile, fixed and strategic energy infrastructures and LNG flows. Regional traditions, the changing dynamics for the supply of natural gas, demand, trade and transport levels in different, remote regions, increasingly influence European natural gas markets and vice versa.

EU energy security is also impacted by geopolitics in the region. This was clearly shown in Russia's move to cut supplies to Ukraine in June 2014 in part due to unpaid gas dues. This compounded the already volatile political relationship between Europe and Russia, and firmly placing the 'original Southern Corridor' initiative back on the European agenda. As such, the 'European Energy Strategy' published in May 2014, ahead of the physical disruptions, set as a priority, accessing more diversified natural gas resources and recognised the importance of the Southern Corridor for reaching supplies from the Caspian region and other countries such as Turkmenistan, Iraq and Iran.[21] The European energy sphere does not currently include the Middle East, and the planned projects intended to bring natural gas from the Caspian region to European markets are marginal in volumes. However, the fear of disruption has led the EU to look at including the Middle East.

Although trade and transport of natural gas via LNG is generally a more flexible option, transboundary natural gas pipelines are the only viable option for connecting the Caspian and the Middle East to Europe. However, given the landlocked geography of the Caspian region, concerns over safety and costs, as well as the geopolitics of the Middle East, render it impracticable. Transport methods and the routes for delivery are limited, with pipelines across Turkish territory appearing to be the most economic route. Therefore, the 'Southern Corridor' concept was initially introduced in 2008[22] to bring together Europe and 'other natural gas producers' in the East by a transit route through Turkey.

---

[21]Commission Staff Working Document, In-depth study of European Energy Security, Accompanying the document, Communication from the Commission to the Council and the European Parliament: European energy security strategy {COM (2014) 330 final}, 2.7.2014.

[22]Second Strategic Energy Review: An EU Energy Security and Solidarity Action Plan {COM (2008) 781}.

The importance of the Southern Corridor has been the promise of linking Azerbaijan gas reserves to Europe and the opportunity for transportation of new volumes from new sources, making possible the further diversification of energy supplies. However, one element of the Southern Corridor—the EU backed Nabucco project which would receive gas via feeder pipelines from Azerbaijan, Turkmenistan, Kazakhstan and Uzbekistan, as well as from Iran, Iraq and Egypt— failed as a result of lack of commitment from producers and financing problems. If constructed, the pipeline would have been the first to operate under the Third Energy Package rules. Instead, a joint project by Azerbaijan and Turkey, the Trans-Anatolian Pipeline Project (TANAP) was announced in December 2011. The pipeline, which will deliver Azeri gas to the EU at Turkey's western border with Greece, is regulated by Turkish law.[23]

In reality, diversification can only be achieved if sufficient transport and import infrastructure capacity is in place and the European gas markets are sufficiently attractive to external suppliers. Nevertheless, uncertainty remained on how to provide the necessary incentives to attract investment into the gas supply infrastructure while at the same time ensuring an integrated competitive market inside the EU.

As mentioned previously, there is currently increasing interplay between different regional gas markets. For most of the twentieth century energy flows were predominantly from East to West. After the Fukushima nuclear disaster in 2011,[24] traditional dynamics began to change owing to increased dependence on LNG in Japan, rapid industrialisation in Asia and LNG available to Europe being diverted to Asian markets where prices are more attractive. The rise of demand in Asia also resulted in the emergence of new pipeline projects destined

---

[23]Intergovernmental Agreement between the Government of Republic of Turkey and the Government of Republic of Azerbaijan Concerning Trans Anatolian Gas Pipeline System, Turkish Official Gazette No: 28592, dated 19.03.2013, Art. 33.

[24]In 2011 due to an earthquake that triggered a tsunami, the nuclear disaster at the Fukushima I Nuclear Power Plant in Japan caused loss of all nuclear capacity which previously provided 26% of power supply and resulted in greater demand for global supply of LNG. Since 2012 Qatar has been the second largest suppliers of LNG to Japan. See Japan Country review available via U.S. Energy Information Administration, available at: http://www.eia.gov/.

to export natural gas from Central Asia further east and the alternative markets that are available to potential suppliers, such as Iran and Iraq in India and Pakistan. Within this fast-changing global market, Europe needed to compete to diversify its supply sources, and market regulation needs to provide appropriate incentives to potential investors.

During the 1990s, most of the national electricity and natural gas markets were still monopolised across most of the EU. Natural gas supply contracts were traditionally long-term to enable large scale infrastructure development through project financing mechanisms. Although this past monopolistic structure stimulated investment in pipeline infrastructure, it was believed that competitive commodity trade *"would not appear spontaneously in this market but had to be created through regulation"*.[25] Hence, driven by the objective of creating competitive gas markets, the EU and its Member States decided to gradually open these markets to competition.

In this way, the EU began to shape its natural gas market structure to respond to what it perceived as detrimental in the structure of the natural gas industry to the EU's regional economic integration ideology. For this, a regulatory approach was introduced which undoubtedly has taken longer than anticipated to complete (in fact, it is arguable whether energy market integration is complete at the time of writing). As such, the EU introduced the first EU wide Gas Directive (first Gas Directive)[26] in 1998. This was replaced by a second set of legislation known as the 'Second Energy Package' encompassing the Gas Directive of 2003 (second Gas Directive)[27] and Regulation 1775[28]

---

[25]This was due to the fact that pipelines are natural monopolies, Dag Harald Claes, The Process of Europeanization—The Case of Norway and the Internal Energy Market (ARENA Working Papers WP 02/12).

[26]Council Directive 98/30/EC of the European Parliament and of the Council of 22 June 1998 concerning common rules for the internal market in natural gas [1998] OJ L204/1 (first Gas Directive).

[27]Council Directive 2003/55/EC of the European Parliament and of the Council of 26 June 2003 concerning common rules for the internal market in natural gas and repealing Directive 98/30/EC [2003] OJ L176/57 (second Gas Directive); Thomas W. Walde (ed.), *The Energy Charter Treaty: An East-West Gateway for Investment and Trade* (1996).

[28]Council Regulation No. 1775/2005 of the European Parliament and of the Council of 28 September 2005 on conditions for access to the natural gas transmission networks [2005] OJ L289/194 (Regulation 1775).

in part because the first Gas directive did not include the desired network access to cross-border transmission and distribution systems. Despite half a decade passing after its adoption, the inherently monopolistic internal natural gas market remained resistant to change, and in response a more vigorous liberalisation came about under the 2009 Third Energy Package. This, first and foremost, expanded the application of internal market rules to external actors, irrespective of the country of origin, such as Russian, Caspian and, prospectively, Middle Eastern producers.

The way internal gas market regulations have evolved since the 1990s represents the EU's ambitious strategy to strip the natural gas industry of its state-controlled and fragmented roots. The Third Energy Package and the EU's two other significant initiatives, the multilateral Energy Charter Treaty (ECT)[29] and subsequent establishment of the Energy Community,[30] prove that the EU's external energy policy is increasingly focused on exporting its rules-based approach to third countries as far as possible.

The combination of these political initiatives focused on further liberalisation. However, in November 2014 a more monopolistic structure has emerged with the concept of a 'European Energy Union' and a new post within the European Commission as Vice-President for the Energy Union.[31] The rationale is to merge resources and negotiation powers within the EU against external suppliers, going as far as to propose the creation of a single European buyer of gas under a single body. Although "*the concept of* [European] *gas market liberalization does not easily syndicated with security of supply*",[32] the creation of an EU-wide gas monopoly was perceived to be at odds with EU competition law

---

[29]Energy Charter Treaty, 2080 UNTS 95; 34 ILM 360 (1995) (ECT) was opened for signature in Lisbon 17.12.1994 and entered into force on 16.04.1998.

[30]The Energy Community was established by an international law treaty in October 2005 in Athens, Greece. The Treaty entered into force in July 2006. For more information visit Energy Community home page available at: http://www.energy-community.org/.

[31]New Heads of the European Union: A New Framework for EU Energy Policy, news release, available at: http://www.foratom.org/newsfeeds.html.

[32]United Nations Economic Commission for Europe (UNECE), Study on Underground Gas Storage in Europe and Central Asia (2013).

and international trade law. Furthermore, this proposal was not appealing for external producers, lenders and non-EU undertakings and it was never clear whether it would enhance supply security. Therefore, to date this idea has not been taken further.

2.3   An Initial Analysis of the EU Natural Gas Market's Attractiveness for Suppliers

Gas imports into the EU have also increased with newly added Member States (although this has not counteracted the rapid decrease in indigenous natural gas production), and at least two thirds of imports originate from outside of Europe, especially from Russia. Furthermore, there is no realistic expectation that domestic shale gas production will change this situation within the next 15–20 years. On the basis of this and set against the most recent security of supply crisis in the aftermath of the 2014 events involving Russia and Ukraine, the need for supply diversification became a priority. During the last five years, however, new dimensions have been added to the equation. Some of the much needed regulation, such as on new capacity allocation has been adopted, bringing an element of predictability for new gas infrastructure investment, with a new proposal in 2018 to allow expansion of the Third Energy Package to offshore pipelines from third countries set out, making the regulatory framework even more complicated. At the same time, the concept of security of supply changed dramatically. The focus is gradually shifting from external events, particularly diversification of supplies to internal question around the future of gas infrastructure vis-a-vis the EU's decarbonisation goals.

The long-term natural gas demand outlook (post 2050) is becoming more uncertain with policies towards the reduction of CO_2 and other GHG emissions. However, the reduction of domestic production and the coal, oil and nuclear phase out, topped with increasing economic growth will require some additional imports of natural gas in the short and medium term. This creates a risk of over-investment in gas and raises the question of whether there is a need to build additional import infrastructure. The drafters of the Third Energy Package gave priority to investments in interconnection pipelines and capacities between Member States and promotion of competition in internal natural gas trade[33] over measures for increasing the attractiveness to Europe for new supply sources. If there is a diversified network in the internal market, even the reliance on a big supplier does not pose significant risks to competition or security of supplies.

The balance of securing supplies and ensuring competition is not a difficult task, but in the past few years the Commission appears to be stuck between the two. The Third Energy Package aimed at increasing competition and diversifying supply sources allows the cheapest gas to enter into the market, but discretion is left to the Commission. The increasing voice of the EC in intergovernmental deals means that it is the Commission and the regulators and not the market who decide. Such uncertainty is a major stumbling block against investment, and at the same time for sustainability and the energy transition.

Construction of adequate gas transportation capacities is an essential element for creating an internal market and interconnectivity with a view to enhancing energy security but at the same time a well-functioning, liquid natural gas market requires sufficient physical gas inside the market. In the view of predicted imports of natural gas into the EU, in the form of both LNG and pipeline gas in the short and mid-term as a result of the rapidly decreasing indigenous production, the attractiveness of the EU market for suppliers and investor requires an assessment. The long term demand is uncertain and any new investment will in principle seek out long term commitments. The decision making depends on the economic viability of projects and the 'rate of return' for capital investments (upstream or midstream). Considering that most non-EU producers are vertically integrated natural gas entities, it also depends on the level of control they have over the entire energy value chain, including transmission networks.

---

[33]"The security of energy supply is an essential element of public security and is therefore inherently connected to the efficient functioning of the internal market in gas and the integration of the isolated gas markets of Member States", Gas Directive, recital 22.

The Third Energy Package introduced a new provision known as the 'Third Country Clause' absent in previous gas directives. This provision requires non-EU undertakings[34] to abide by internal market rules, including tariff regulation, and to meet the terms of unbundling provisions and permit 'third party access' to their transmission networks should they wish to acquire control over an EU transmission system operators.[35] This meant loss of control and revenues for the suppliers. The Third Energy Package and subsequent efforts by the Commission to increase its control over investment decisions therefore constituted a concern for producers and a significant de-stimuli for financial investors. However, the legislation provides exemptions and derogations from the rules on unbundling,[36] third party access and tariff regulations to promote investment in infrastructure, suggesting it is impracticable to build a new project or expand existing capacity without the inclusion of some formal limits to the legal framework for liberalisation.

The exemption decisions are taken on a case-by-case basis by respective National Regulatory Authorities (NRA), and the EC has a right to veto the decision. This approach may not be ideal for the international investment community. Until the process of decision making in the energy sector in the EU is legislated in a way that is transparent and accessible to stakeholders the exemption procedure becomes political to the detriment of the financing of prospective transboundary pipeline projects. This then required many gas infrastructure projects to be financed through public funds.

Furthermore, the EU's external energy strategy is increasingly focused on exporting its rules-based approach to third countries and expanding market integration. In this regard, as per Article 4 of the Treaty on the European Union (TEU),[37] all energy related

---

[34]Preamble 21, Gas Directive.

[35]Art. 32, Gas Directive.

[36]Art. 36, Gas Directive.

[37]Article 4 requires Member States to take all appropriate measures to ensure fulfilment of the obligations arising out of the Treaties or resulting from the acts of the Union institutions, Treaty on European Union (Maastricht text), 29 July 1992, 1992 OJ C191/1 (Maastricht TEU).

intergovernmental agreements (IGAs) signed with third countries are required to be in compliance with EU legislation, in particular the Third Energy Package and competition law. To be compliant, all new and existing IGAs need to be notified to the Commission.[38] The Commission subsequently ensures that all such agreements and all infrastructure projects within the EU territory fully comply with the EU *acquis*.[39]

Even if the IGAs in question are not notified to the Commission at an earlier stage, the Commission intervenes *ex facto*. For instance, the Commission called in the South Stream project's IGAs between Russia and Member States for renegotiation as the agreements were thought to violate internal market rules and competition law. The South Stream project however would not bring new sources of supplies to the EU; it promised instead to provide a new direct route for transport existing supplies to minimise transit risks along the Ukrainian transport corridor. In other words, there is an increasing involvement of political bodies in the decision making for new cross-border pipeline infrastructure development and also for LNG facilities. The impact of this on the future of gas infrastructure in the EU will be discussed in the chapters below. It is important at this point to present the experience in project development in the natural gas sector, in particular with pipelines as the lessons learned will be instrumental in assessing the future of renewable and decarbonised gas investments, some of which will involve the use of existing gas pipelines and there will also be a need to import new sources of green gas from third countries.

---

[38]Decision No. 994/2012/EU of The European Parliament and of the Council of 25 October 2012 establishing an information exchange mechanism with regard to intergovernmental agreements between Member States and third countries in the field of energy, recital 1, 2, 9 and 11.

[39]Regulation (EU) 2017/1938 of the European Parliament and of the Council of 25 October 2017 concerning measures to safeguard the security of gas supply and repealing Regulation (EU) No. 994/2010. (Text with EEA relevance.)

2.4   How Are Pipeline Projects Developed?

There are four alternatives for developing a cross-border import pipeline to deliver gas to the EU.[40] First, the pipeline might be connected to an external EU border, such as the TANAP, which will transport natural gas from Azerbaijan's Shah Deniz II field through Turkey to its western border with Greece, connecting the Trans-Adriatic Pipeline (TAP) running from Greece and Albania to West of the Adriatic coast in 2020.[41] In this case, the Third Energy Package will not be directly applicable to the TANAP pipeline. However, the project owner, Azerbaijan's state-owned company the State Oil Company of Azerbaijan Republic (SOCAR), has interest in delivering gas to the European end-users. As such it applied to acquire control of the Greek TSO DEFSA.[42] In order to operate the Greek transmission system SOCAR had to comply with the requirements of Article 11 of the Gas Directive which relates to the certification of TSOs controlled by a person from a third country (Third Country Clause).

An additional condition, applied in respect of third countries, is that investment by third country entities should not endanger the security of supply of the Member State(s) in question or of the Community. In such a case, the Member State has the ultimate right to refuse requests for certification.[43] The EC initiated an investigation against DEFSA's privatisation under competition law and raised concerns that, in light of "*SOCAR's participation in the Shah Deniz II field and in the TANAP pipeline, SOCAR could use its control over DESFA's network to foreclose gas supplies from other sources*".[44] The deal collapsed in November

---

[40]Conclusions of the 6th EU–Russia Gas Advisory Council Vienna, 29 January 2013.

[41]Azerbaijan started first gas deliveries to Turkey via TANAP on 30 June 2018. The first gas deliveries to Europe via TAP are expected to start in 2020.

[42]Azerbaijan's SOCAR Signs Deal to Buy 66% of Greece's Gas Grid Operator DESFA, Platts, 23.12.2013, available at: http://www.platts.com/latest-news/natural-gas/moscow/azerbaijans-socar-signs-deal-to-buy-66-of-greeces-26569771.

[43]Art. 11(5)b, Gas Directive.

[44]Commission Opinion of 17.10.2014 correcting Opinion C(2014) 5483 final of 28 July 2014 pursuant to Article 3(1) of Regulation (EC) No. 715/2009 and Articles 10(6) and 11(6) of Directive 2009/73/EC—Greece—Certification of DESFA, C(2014) 7734 final.

2016, after the EC banned the sale of the controlling stake in DESFA, demanding a reduction of the SOCAR stake to 49%.

Second, project investors or the producers might opt for delivering gas into the EU territory in compliance with the Third Energy Package and without any derogation,[45] such as the Gazprom led, now defunct South Stream pipeline. This project was planned to traverse, alongside a number of EU countries, members of the Energy Community which makes compliance with the EU energy legislation obligatory. The Russian state-owned company chose not to apply for an exemption decision but instead sought to find a level playing field within the Third Energy Package regime via its negotiations with European counterparts, such as through the Gas Advisory Council Meeting and Russia–Europe dialogue platform.

As a third possibility, a supplier may bring the gas inside the EU based on exemptions from the Third Energy Package. The exemptions are given for a defined period of time and could be settled as full or partial. For instance, the OPAL[46] and Gazelle[47] pipeline sections of the North Stream pipeline[48] were granted exemption from only certain provisions of the then applicable second Gas Directive. Gazelle is exempted from 'regulated third party access' and from the 'regulation of tariffs' for a period of 23 years, of a maximum of 30 bcm of direct forward-flow capacity of the pipeline. The OPAL pipeline is exempted from network access and transit fees regulation for 22 years by the decision of Germany's NRA. However, the Commission interfered with the decision that allowed OPAL to run at full capacity and limited the exemption to 50% of OPAL's capacity. The other half of the facilities' capacity

---

[45]Art. 36, Gas Directive.

[46]The OPAL gas pipeline is a ground-based continuation of the North Stream gas pipeline, which supplies Russian gas to Western Europe via the Baltic Sea bed. For factual information visit OPAL pipeline's home page available at: http://www.opal-gastransport.de/home.html?L=1.

[47]The 'Gazelle' pipeline crosses the Czech Republic and connects the OPAL pipeline in Eastern Germany with the MEGAL pipeline system in Southern Germany.

[48]Factual information relating to the Nord Stream project is available at: http://www.nordstream.com/en.

are to be offered to third-party companies in Czech markets.[49] By virtue of this 50% restriction on the throughput capacity, the investors could not pay back the sunk costs efficiently. Natural gas projects have a specific cost structure. Production and transport of natural gas requires very large projects with high fixed capital costs and relatively low operation cost, thereby requiring full utilisation of capacity (maximum flow relative to design capacity) to justify the project by maximum revenues. Where the project is not operated at full capacity the already lengthy pay-back period is prolonged at a loss to the operator.

The fourth option is to supply the EU territory on the basis of the Projects of Community Interest (PCI) regime.[50] Alternatively, the project could be granted Project of Mutual Interest (PMI) status. PCIs are those identified by the EC as key infrastructure projects in 2013, which in its opinion *"contribute to the development of the trans-European transport network through the creation of new transport infrastructure, the maintenance, rehabilitation and upgrading of existing transport infrastructure and through measures promoting its resource-efficient use"*.[51]

PCIs benefit from accelerated permitting procedures and are eligible for financial aid from the instruments of the EU. For instance, the TAP connecting Caspian resources via the TANAP pipeline is granted a PCI status. This facilitated an exemption decision for the project as the EU acknowledged its contribution to diversification and security of energy supply. The TAP pipeline is the first project to be developed under an exemption from the Third Energy Package. The exemption is given for

---

[49]Commission Decision of 20 May 2011 on the exemption of the 'Gazelle' interconnector according to Article 36 of Directive 2009/73/EC; the Decision of the German Regulatory Authority of 2011 are available at: http://ec.europa.eu/energy/infrastructure/exemptions/doc/exemption_decisions.pdf.

[50]The European Commission, together with European Member States, is deciding on which energy projects should be given the status of a Project of Common Interest (PCI) at the EU level and thus be subject to preferential treatment in the Member States and benefit from faster and more efficient permitting procedures. Such projects are considered to be of public interest from an energy policy perspective and can under certain conditions be also eligible for EU co-financing.

[51]Proposal for a regulation of the European Parliament and of the Council on Union guidelines for the development of the Trans-European Transport Network Brussels, 19.10.2011 {COM (2011) 650 final}, Art. 7.

25 years (corresponding to the gas supply agreements) from requirements on third party access, tariff regulation and ownership unbundling (OU)[52] for the initial 10 bcm capacity. There are however some conditions the Commission imposed on the project investors.[53] Furthermore, the third party access exemption excludes at least 5% of capacity for short-term contracts with a duration of less than 1 year (inclusive) and 10% of actually build expansion capacity.[54]

A PMI is "*a project involving both the Union and one or more third countries* [covered by the European Neighbourhood Policy,[55] the Enlargement Policy,[56] the European Economic Area and the European Free Trade Association[57]] *which aims to connect the trans-European transport network with the transport infrastructure networks of those countries to facilitate major transport flows*".[58] The EU may support a PMI based on existing or establishing new coordination and financial instruments.[59]

Existing capacities in non-exempt natural gas pipeline projects, or after the duration of derogations, are by default allocated through auctions as of 1 November 2015, the application date under the Capacity Allocation Mechanisms Network Code (CAM NC).[60] Initially, the Third Energy Package did not provide a procedure for the construction and utilisation of new pipeline capacity. However, an 'Incremental Capacity'

---

[52]Art. 9, 32, 41 (6), 41 (8) and 41 (10), Gas Directive.

[53]Commission Decision of 16.5.2013 on the exemption of the Trans Adriatic Pipeline from the requirements on third party access, tariff regulation and ownership unbundling laid down in Articles 9, 32, 41(6), 41(8) and 41(10) of Directive 2009/73/EC, para. 4.1.

[54]TAP Exemption Decision, para. 4.1.10.

[55]Communication from the Commission, European Neighborhood Policy Strategy Paper {COM (2004) 373 final}, 12.5.2004.

[56]As of November 2014 this includes Albania, Bosnia and Herzegovina, Iceland, Kosovo, Montenegro, Serbia, The Former Yugoslav Republic of Macedonia, and Turkey. More information available via: http://ec.europa.eu/enlargement/.

[57]The European Economic Area (EEA) unites the EU Member States and the three EEA EFTA States (Iceland, Liechtenstein, and Norway) into an Internal Market governed by the same basic rules.

[58]New TEN-T Guidelines proposal {COM (2011) 650}, Art. 3(b).

[59]New TEN-T Guidelines proposal {COM (2011) 650}, Art. 8.

[60]Commission Regulation (EU) No. 984/2013 of 14 October 2013 establishing a Network Code on Capacity Allocation Mechanisms in Gas Transmission Systems and supplementing Regulation (EC) No. 715/2009 of the European Parliament and of the Council, OJ L273.

chapter has been adopted in the framework of the Regulation (EU) 2017/459 of 16 March 2017 establishing a network code on capacity allocation mechanisms in gas transmission systems and repealing Regulation (EU) No. 984/2013 (CAM NC) which applies as of 6 April 2017.[61] The amendment offers an 'Alternative Allocation Mechanisms' procedure, as an alternative to the default auction method. While the Network Code does not define the process as an 'Open Season' it carries the characteristics of a classic two-step open season procedure.[62] First, it enables identification of actual market demand and second to allocate capacity on a transparent and non-discriminatory basis accordingly. This change to the regulation was necessary as allocation of capacity based on auctions is not financeable for large scale projects such as Nabucco, TAP, South Stream and North Stream. New capacity is the result of the evolution in the demand for capacity and available capacity. A market based, investment friendly, Open Season/Alternative Allocation Mechanisms for new cross-border projects is the only economically viable way to construct major pipelines in Europe unless the project is exempt from the Third Energy Package (or the project is listed as a PCI or PMI).

The CAM NC stipulates that irrespective of the duration of the supply contract, capacity can only be booked for 15 years in the EU.[63] The rationale is that the supplier would book/block the capacity for so long that it would foreclose short-term markets for small competitors. Under the default auction procedure of CAM NC, a network user can book yearly, quarterly, monthly, daily and within-day standard capacity products. The TSOs are called to provide details on the available interconnection capacity between 'Entry-Exit' zones (so-called 'hubs')[64] and shippers to submit bids and offers for cross-border gas trades

---

[61]Commission Regulation (EU) 2017/459 of 16 March 2017 establishing a network code on capacity allocation mechanisms in gas transmission systems and repealing Regulation (EU) No. 984/2013, OJ L72/1 (CAM NC).

[62]Katja Yafimava, *Building New Gas Transportation Infrastructure in the EU—What Are the Rules of the Game?* OIES 2018.

[63]CAM NC, Art. 11(3).

[64]Regulation 1775, Art. 18(3).

with capacity implicitly allocated (capacity and commodity are traded together) to trades via one or a limited number of joint web-based booking platforms operated by TSOs or third parties.[65]

The abovementioned system prescribes the way energy is to be transported between and across borders in Europe, namely the 'Entry-Exit' system which became obligatory as of September 2011.[66] Thereby, the previously applied method of transport, the 'Point-to-Point' system, was abandoned. Under the Point-to-Point system gas used to flow through a predetermined route, the tariffs were distance based and capacity was commonly allocated on a first come first served basis.

In the Entry–Exit system transmission tariffs do not follow contractual paths[67] and capacity is 'freely allocable'. In other words, natural gas brought into the European pool system, at any entry point, can be made available for off-take at any exit point within the system on a fully independent basis. Similarly, this has the effect that each exit point can be considered as being supplied from any entry point without any restrictions. Another feature of the Entry–Exit system is that between each entry and exit zone, virtual transportation zones are created where ownership of the capacity may be traded in the short term. Despite different implications, the Entry–Exit system exists between different Member States and it is applicable in the EU at large.

It is important to note that the Entry–Exit system is a product of the European 'Gas Target Model' (GTM) initiated by the Council of European Energy Regulators (CEER) in 2010 and established by the Commission and EU Energy Regulators. The GTM is a vision of how the regulatory design of a single European gas market and gas destinations into Europe should look. Transformations of this magnitude naturally brought uncertainty and constituted some concern for gas

---

[65]In April 2013 major European TSOs from Austria, Belgium, Denmark, Germany, France, Italy and the Netherlands established a joint European capacity platform called PRISMA to implement Regulation on Capacity Allocation Mechanisms.

[66]Regulation 1775, recital 19 and Art. 13(1).

[67]The network code on rules regarding harmonised transmission tariff structures for gas (TAR NC) is the Commission Regulation (EU) 2017/460 of 16 March 2017 establishing a network code on harmonised transmission tariff structures for gas.

stakeholders, including, and perhaps predominantly, third country suppliers and investors.[68] Although the GTM acknowledges the importance of providing the right signals for investment in the cross-border network infrastructure and for long-term trading of gas, the new regulatory framework initially brought threats to network users, such as the risk of contractual mismatch mentioned above.

Another criticism was that the Entry–Exit system also increases transaction costs, as network users have to pay transmission tariffs for each capacity booking at individual entry and exit zones.[69] Tariffs are the revenue received by the TSOs for the sale of capacities. The Third Energy Package does not impose any standardised tariff structure on Member States and the NRAs have the responsibility for fixing or approving tariffs or methods for their calculation.[70] In an effort to harmonise transmission tariff structures and to allocate costs for projects within the regulatory zone of the EU, a Network Code on Tariffs (TAR NC) have been developed. TAR NC now foresees that the payable price for capacity products should be calculated as the sum of its reserve price and, if any, the auction premium. The Framework Guidelines from which TAR NC is developed suggest that the reserve price should be identified as a floating price.[71] Floating tariffs imply that where the revenues of the TSO are less than expected, the deficiency should be compensated for via a tariff increase for the following year. This creates a situation where the reserve price for capacity that has already been booked could increase as the final tariff is not set until after the auctions have taken

---

[68]That said, there are still certain separate transit systems within the EU such as Bulgaria and Romania where no third-party access is guaranteed. Furthermore, the Yamal pipeline system in Poland uses a peculiar entry exist system covering only one large trunk pipeline with no virtual points, see Study on Entry–Exit Regimes in Gas, Part A: Implementation of Entry–Exit Systems, DNV KEMA, by order of the European Commission—DG ENERGY, 11.12.2013, available at: https://ec.europa.eu/energy/sites/ener/files/documents/201307-entry-exit-regimes-in-gas-parta.pdf.

[69]Council Regulation No. 715/2009 of 13 July 2009 on conditions for access to the natural gas transmission networks and repealing Regulation (EC) No. 1775/2005 [2009] OJ L211/36 (Gas Regulation), Art. 13.

[70]Recital 32, Gas Directive.

[71]Framework Guidelines on rules regarding harmonised transmission tariff structures for gas 29 November 2013.

place. In 2017, radical changes to this system were proposed in policy discussions. In the Madrid Forum in 2017, a proposal was made to remove all the internal tariffs and move all gas entry points to the border. ACER also noted in 2019 that they expect concerns over gas tariffs to grow as more and more long-term contracts expire and they will be replaced with shorter term contracts. As short-term tariffs are higher than the long-terms ones there is a risk that TSOs will not be able to recover their revenues and this could result in barriers to trade.[72] The 2020 gas package, in addition to addressing the regulatory requirements regarding low carbon and renewable gases, could be expected to target an improvement of these concerns with the existing GTM.

Furthermore, in an Entry–Exit regime, the capacity can be offered as either 'bundled capacity' or 'unbundled capacity'. Bundled capacity means a corresponding entry and exit capacity on a firm basis at both sides of every interconnection point.[73] Although the TSOs have a duty to maximise the quantity of bundled capacity,[74] capacity on each entry and exist zone may not match in all cases,[75] resulting in contractual or physical congestion risks.[76] As such, out of over 350 interconnection points in Europe, at least 118 were reportedly congested in 2013.[77]

Contractual congestion occurs when there is unutilised remaining capacity and, against the prevailing demand from network users for capacity, the TSO cannot allow more injections into the system as the capacity is fully booked by existing contracts. Since, in the Entry–Exit system, the capacity may leave at any exit point, a TSO needs to

---

[72]ACER Public Consultation on the Bridge Beyond 2025, PC_2019_G_06, Consultation period: 23 July 2019–1 September 2019.

[73]CAM NC, Art. 3.4.

[74]CAM NC, Art. 6.

[75]The European Federation of Energy Traders (EFET), An EFET Position Paper: Advancing the EU Internal Energy Market: Sector Priorities for the Juncker Commission, 12.11.2014, available at: http://www.efet.org/.

[76]Article 2 of Regulation 715 defines them as follows: *"Contractual congestion, means a situation where the level of firm capacity demand exceeds the technical capacity* [and] *Physical congestion, means a situation where the level of demand for actual deliveries exceeds the technical capacity at some point in time."*

[77]ACER annual report on contractual congestion at interconnection points (Period covered: Q4/2013), para. 37.

calculate and reserve the commercial capacity in a network, taking into account all exit points in the system.

Physical congestion occurs when, irrespective of demand for capacity, no additional actual flows can be accommodated via a transmission network because the maximum technical capacity is reached. The obvious solution for physical congestion is to attract further investment into new or incremental capacity. Physical congestion provides a signal for investment, whereas contractual congestion indicates that the network is not efficiently used. Acknowledging congestion risks, in 2014 a Congestion Management Mechanism was attached to Regulation 715. It stipulates that if capacity is not used it could be given to secondary markets,[78] despite the fact that secondary markets for transport capacity did not previously exist in all Member States. In 2017, based on the ENTSOG data sets and Booking platform data, only less than 7% (17) of the 262 interconnection point sides were contractually congested. This is as a direct result of successful implementation of the Congestion Management Mechanism.[79] In order to avoid congestion, ACER recommends, among others, that congestion management procedures are applied as a preventive measure before contractual congestion occurs.

In order to minimise congestion problems, the TSO needs to undertake an effective balancing mechanism to offset the gap between commercial and the physical gas networks. In this regard, a Network Code on Gas Balancing of Transmission Networks (BAL NC) was published in 2014, applicable as of October 2015.[80] BAL NC sets out a vision of market-based balancing at a single virtual trading point inside an entry–exit system.

Due to a range of complex interactions in the changing EU energy market, long-term market uncertainties carry the risk of shippers being reluctant to invest in long-term capacity bookings. This is of course

---

[78]Regulation 715, Annex I, guidelines on congestion-management procedures in the event of contractual congestion, para. 2.2.

[79]ACER annual report on contractual congestion at interconnection points (Period covered: 2017).

[80]Commission Regulation (EU) No. 312/2014 of 26 March 2014 establishing a Network Code on Gas Balancing of Transmission Networks Text with EEA relevance, OJ L91.

exacerbated by long-term demand uncertainties as a result of the EU Member States' commitments to the Paris Climate Agreement. While short-term capacity bookings and supply contracts increase transparency and flexibility in mature markets, long-term transport and supply contracts and capacity bookings have been fundamental for the development of new major pipelines and LNG facilities as they provide bankability and distribute risks between gas producers and importers.[81] Furthermore, the lack of long-term bookings may create a misleading assumption in the market—such that there is no need for capacity. Since the level of market maturity is uneven throughout Europe, such that sufficient import capacities and liquid hubs are already in place in North-Western and increasingly in Central Europe but absent in South Eastern Europe, a 'one size fits all approach' may not enhance supply security per se. In the view of the Commission, the absence of liquidity in southern Europe discourages alternative suppliers.[82]

Considering the degree of uncertainty in the potential proliferation of transaction costs and the bankability of natural gas projects, there is a factual risk of a decline in future upstream and midstream investment incentives for Europe, while it is still needed in the short and medium term.

There is no incentive in the Network Codes for those shippers who are willing to pay for the construction of the infrastructure and full 100% capacity of the pipeline is not given to the pipeline owners anymore. In principle, if there is no market guarantee for around 20 years and if gas cannot freely flow no company would be willing to build pipelines. Although no new major natural gas pipelines may be necessary, upstream investments are still needed in order to cover the import needs until at least 2050, when scalable hydrogen should be expected to be available as a substitution. As an all-electric future even with advancements in battery storage, while ideal for addressing climate change, is not achievable when taking into account the magnitude of

---

[81]Draft Vision for a European Gas Target Model: A CEER Public Consultation Paper, p. 8.

[82]Commission Staff Working Document, In-depth study of European Energy Security, Accompanying the document, Communication from the Commission to the Council and the European Parliament: European energy security strategy {COM (2014) 330 final}, 2.7.2014.

gas demand in the EU, particularly from the hard to decarbonise sectors of the economy such as the energy intensive industries.

Furthermore, third country producers perceive natural gas markets from a 'resource policy' perspective which is significantly different to the 'energy policy' perspective of their European counterparts.[83] Resource policy implies an inherent tendency to tightly control the entire energy value chain through state ownership of the export pipeline infrastructure. This is to ensure that the energy sector contributes to national developmental goals and that natural resources are developed to maximise resource rent. In the case of natural gas exports via pipelines, where specific production is linked with a specific market, sales from the commodity is the principal revenue stream for the upstream investment project. It is for this reason that upstream investors traditionally had interests in both mid-stream and downstream segments of the energy value chain (such as SOCAR and Gazprom).

This approach, however, might not be acceptable to the host state where the pipeline is located, specifically if it pursues a liberalised market model. As such, unless a project is granted an exemption, in order to receive a certificate to operate a pipeline in the EU any natural gas undertaking must satisfy one of the three types of the unbundling model: the OU, the independent system operator (ISO) or the independent transmission operator (ITO).[84] OU means that a vertically integrated production or supply undertaking may not own or retain decisive control over the transmission operations. Under the ISO system, the owner of supply or production activities upstream may not, at the same time, operate the transmission system. Although the upstream undertakings may own the transmission infrastructure assets, the system operator itself must be a separate entity fully independent from such producers or suppliers. The ITO system permits the transmission operator to be the same entity as producer/supply undertaking and retaining ownership. However, a stringent set of regulations are enforced in order

---

[83]Kim Talus, United States Natural Gas Markets, Contracts and Risks: What Lessons for the European Union and Asia-Pacific Natural Gas Markets? *Energy Policy*, Volume 74 (2014), p. 33.

[84]Art. 9 and 11, Gas Directive.

to ensure that an ITO is managed independently, supervised adequately and that sufficient deterrent fees are charged for any non-compliance. Member States have the right to determine the type of OU system applicable in its internal market.

As a method of financing, a combination of sponsor equity and project financing has been traditionally used in the oil and gas industry since the nineteenth century. In project financing, a major part of investment funds for the project is raised on capital markets rather than through internal equity. Typically, project finance is favoured when a long financing period, often over 10–25 years, is required due to the time involved in construction and operation, hence the need for a long revenue stream period for the project. A special purpose project entity known as a Special Purpose Vehicle (SPV) is incorporated to solely carry out the project. The lenders have limited or no recourse to the project sponsors for repayments and their only guarantee is the cash flows and assets of the SPV in question. The assets of the SPV are often worth less than the loan provided and the loan: equity ratio is generally as high as about 70:30 for oil and gas projects.

Project finance is attractive for lenders as it provides a high level of leverage. However, lenders often undertake a detailed 'due-diligence' procedure before providing any loans. The banks would only assume the risks that are measurable such as market risks which implies volume and price risks. Commercial risks of this type can be mitigated between borrowers and lenders via long-term contracts and 'take or pay' clauses under which the buyer pledges to take or otherwise pay for a minimum quantity of gas per year.

As mentioned above, specific production and specific consumers are linked in network-bound gas projects. Volumes cannot be easily diverted and given that the project is an immobile fixed infrastructure, the investment cannot be relocated. This means that if there is no long-term commitment from the buyer adequate to provide a rate of return for upstream and mid-stream investments, the project will not be realised. For instance, in the TANAP project, all Shah Deniz (upstream) contracts with different off-takers are long-term.[85] Similarly in Europe, capital intensive major natural gas infrastructure projects have traditionally been constructed based on long-term commitments (around 15–25 years) with 'take-or-pay' clauses.

However, the Commission perceives long-term contracts as detrimental to competition due to risk of market foreclosure. Accordingly, the Commission increasingly promotes short-term contracts with durations below 5 years. Since the early 2000s, inspired by the UK and US gas markets, the main trend has been towards gas-to-gas competition with prices linked to spot prices at European gas trading hubs. Thus, some of the Shah Deniz contracts are indexed to the Netherland's Title Transfer Facility (TTF),[86] while others remain oil-indexed. Oil or replacement value indexation has been used as a pricing formula for natural gas over the past 40 years across the globe (except for the US where replacement value indexation has never been the case) given that natural gas was, initially, intended to be a substitute for other fossil fuels such as coal and petroleum. Indeed, when exploitation in the Groningen field in the Netherlands began, natural gas was expected to take over the share of oil supply due to its promise of lower carbon emissions and higher efficiency.

Today oil and gas are not competing fuels which renders oil-linkage somewhat irrelevant. However, a correlation between oil and gas prices exists in the long term. In the short term the price of these commodities can diverge as gas is more exposed to seasonality. Accordingly, the Third Energy Package and GTM aim to accelerate the emergence of virtual trading hubs within the EU. The rationale is to achieve efficient price signals, based on supply and demand to decrease prices to benefit consumers. Indeed, the hub prices have been lower than oil-indexed prices under long-term supply agreements for most of the last decade.[87]

---

[85]International Energy Agency, *World Energy Outlook, 2014*, p. 4, available at: https://www.iea.org/publications/freepublications/publication/WEO2014.pdf.

[86]International Energy Agency, *World Energy Outlook, 2014*, p. 4, available at: https://www.iea.org/publications/freepublications/publication/WEO2014.pdf.

[87]Commission Staff Working Document Accompanying the Document Report from the Commission to the European Parliament, the Council, the European Economic and Social

It could be concluded that today oil-indexation is losing ground in the European market but continues to play an important role in certain regions, such as the Mediterranean, Southeast Europe and the Baltics.

2.5   Chapter Conclusions

This chapter concludes that in principle, where there is a functioning, competitive market and predictable demand to guarantee return of their investment for project sponsors, the finance needs for any future infrastructure should be born by the private sector. Considering the risk of stranded assets, this would be ideal to let the private sector decide where and when to invest along the energy value chain. The investment needs in the electricity and gas sectors were set significantly high in 2014, near €1 trillion towards 2020.[88] It was considered then that around €70 billion needed to be invested in the gas industry, including construction of import facilities.[89] Most of this €70 billion was expected to come from the private sector and the voices from inside the Commission warned at the time that *"investment volumes, needed for infrastructure, are likely to go beyond the financing ability of many TSOs"*.[90] Only an amount of €5.85 billion was been approved for direct EU support for PCI projects for the period 2014–2020 through public sources such as the Connecting Europe Facility, European Bank of Reconstruction and Development (EBRD) and the European Investment Bank (EIB). Therefore, most of this money will have to be raised through loan financing mechanisms, which require a business case.

Committee and the Committee of the Regions Energy Prices and Costs in Europe, Brussels, 9.1.2019 SWD (2019) 1 final, Part 1/11.

[88]Based on data from PRIMES, ENTSO-E, KEMA and ECOFYS.

[89]Trans-European Energy Infrastructures: The New Guidelines and the Connecting Europe Facility, Policy Officer European Commission Kitti Nyitrai, DG Energy, Internal Market I: Networks and Regional Initiatives.

[90]Directorate General for Energy, Klaus-Dieter Borchardt, Towards a single energy market—required investment in gas and electricity infrastructure (European Forum for Science and Industry, 18 December 2013).

3

The Energy Mix: Decarbonisation and Natural Gas in EU Energy Policy

3.1 Introduction

As an introduction to this chapter, it is important to reiterate that natural gas, as a lower GHG emissions substitute to coal and oil, and alongside renewable energy, plays an important role as a transition and grid-balancing fuel in combatting the effects of climate change. The renewable energy sector needs natural gas at least until large scale electricity storage becomes viable. Natural gas also has the advantage of flexible transportation via pipelines, as well as LNG to densely-populated areas and liquefied petroleum gas to rural areas, alongside small-scale and community-owned renewables. Yet, it has not been utilised to its full potential since the financial crisis, topped with competition from coal.

EU natural gas demand projections are highly unpredictable. For instance, Exxon Mobile estimates demand to reach almost 600 bcm by 2035, the US Energy Information Administration (EIA) predicts even higher demand at 716 bcm by 2040. The IEA, on the other hand, forecasts demand to reach just under 500 bcm in 2030. Future demand projections are influenced by a number of factors, such as technological

© The Author(s) 2020
G. Mete, *Energy Transitions and the Future of Gas in the EU*, Energy, Climate and the Environment, https://doi.org/10.1007/978-3-030-32614-2_3

innovations, for instance in CCUS technology, the effectiveness of EU legislation, e.g. how the EU Emissions Trading Scheme (ETS) will play out. Demand is also impacted by prospects for economic growth and policy directions on nuclear power and the environment and on changes in commodity prices. In all circumstances a sharp decrease in indigenous production will mean that demand for natural gas imports will increase towards 2050 for heat, transport and power generation.

The EU decarbonisation agenda has also been shaping the future of gas demand in Europe. It is a three-pillar policy, which includes increasing the share of renewable energy resources, lowering carbon dioxide (CO_2) and other GHG emissions and improving energy efficiency. The latest Clean Energy Package, introduced in November 2016, confirms that natural gas is an important enabling fuel towards a low carbon economy. The Clean Energy for all Europeans package consists of eight legislative acts. In March 2019, the European Parliament completed the parliamentary approval of the new Electricity market Regulation and Electricity market Directive as well as of the Regulations on Risk Preparedness and on ACER. The Regulations enter into force immediately with a date of application of 1 January 2020 for the Electricity Regulation and the Electricity Directive will have to be transposed into national law within 18 months. The Governance of the Energy Union Regulation the revised Energy Efficiency Directive, the revised Renewable Energy Directive and the Energy Performance of Buildings Directive have already entered into force in 2018. In short, it underlines EU leadership in tackling global warming and provides an important contribution to the EU's long-term strategy of achieving carbon neutrality by 2050. However, experts commented that the Clean Energy Package seems to be dissatisfied with the liberalisation of markets that came with Third Energy Package, allowing markets to decide on sources for supplies, limiting diversification of supplies, as most options for new sources of supplies are, for a number of reasons, expensive.[1] This will be discussed subsequently in the next two chapters.

---

[1] Tim Boersma, What's Next for Europe's Natural Gas Market? Brookings, 15.03.2016, available at: https://www.brookings.edu/blog/order-from-chaos/2016/03/15/whats-next-for-europes-natural-gas-market/.

Decarbonisation is taken seriously by the Commission, as the block remains the world's third largest emitter of CO_2 after China and the US.[2] The EU took a lead in the climate change negotiations and despite acknowledging the role natural gas plays in lowering CO_2 emissions, by replacing coal and oil in transport via hydrogen (see more below), open support for natural gas has not been forthcoming by the Commission. The dialogue on natural gas is becoming increasingly politicised, causing disparity and a lack of meaningful communication between the industry, the Commission and climate change lobbies.

Irrespective of open backing for natural gas by the Commission, it will play an important role in the EU energy mix towards 2050, particularly in transport, heating and electricity sectors and as a back-up for intermittent renewables. This chapter will look into market dynamics in the EU, including consumption, production trends and the current energy mix. In doing so this chapter seeks to map out existing pipeline projects with external suppliers and LNG import terminals. This analysis, which also includes a study of gas prices, unconventional potential and gas storage capacity in the EU, aims to demonstrate the important role natural gas infrastructure plays in the context of EU energy security. This chapter will also analyse the impact of decarbonisation policies on natural gas demand and scrutinises whether further investments in natural gas infrastructure in the future will be required. Furthermore, it will discuss how different fuels contribute to the energy mix, their CO_2 equivalents, and the status quo of technological advancements and investment levels on hydrogen, synthetic gas, biomethane and around carbon storage, its transportation and utilisation. This inevitably involves discussion on planned decommissioning activities. This chapter will conclude with initial thoughts on the 2020 gas package; these will further be discussed in the following chapters, and further recommendation will be drawn in Chapter 6 on the decision-making framework.

---

[2] Global Greenhouse Gas Emissions Data, United States Environmental Protection Agency, available at: https://www.epa.gov/ghgemissions/global-greenhouse-gas-emissions-data.

3.1.1 A Reality Check

There is a lot of misleading information on the status of gas markets not only in the EU but at the global level. As the EU is part of an increasingly interconnected gas market, it is important to provide a reality check before diving into further detail of the EU energy mix. First, in contrast to what many claim, there is no gas supply scarcity in the global market. US shale gas production, and tight shale to lesser extent, have completely over-turned conventional wisdom on the import and export outlook for the US. In addition to the US, there are new supply sources including from the Arctic and the Caspian regions and potential supplies from Mozambique, Romania, Egypt, Senegal, Cyprus, Mauritania, Argentina, Canada (East-Coast) and the Shetland Islands (Scotland).

Secondly, gas demand in the EU is not declining sharply despite improvements in energy efficiency and substantive renewable energy deployment. As mentioned in the introductory chapter, gas imports to the EU actually increased by 6.5% as a result of declining production. Demand decreased compared to 2014 levels but has since recovered. There may be a further increase as a result of gas use in LNG bunkering and as a fuel for large scale tracks. There are already new projects such as Gasunie in Finland that uses LNG. International Maritime Organisation (IMO) Sulphur Dioxide (SO_x) standards will come into force in 2020 and will make LNG as a fuel in maritime sector even more important.[3] Limiting SO_x emissions from ships improves air quality and protects the environment. Gas power plants were struggling in the EU because of cheap coal but the increase in carbon pricing now serves to help with coal to gas switching. Of course, there has been also

---

[3]IMO regulations to reduce SO_x emissions from ships first came into force in 2005 under Annex VI of the International Convention for the Prevention of Pollution from Ships (known as the MARPOL Convention). Since then, limits on SO_x have been progressively tightened. From 1 January 2020, the limit for sulphur in fuel oil used on board ships operating outside designated emission control areas will be reduced to 0.50% m/m (mass by mass). This will significantly reduce the amount of SO_x emanating from ships and should have major health and environmental benefits for the world, particularly populations living close to ports and coasts.

a nuclear switch off in some EU countries. Therefore, demand growth is foreseen going forward (this will be discussed further below).

Another misperception is that gas trade is based on long-term deals. Whereas actually one third of LNG is now traded as sport. Portfolio buyers are using LNG for short term for arbitrage. There is a difference between short-term and spot contracts. Short-term contracts are those with less than 4 years and spot contracts are those volumes with 3 months. Long-term contracts are still being signed on LNG such as by China, Thailand and Poland to overcome credit challenges. Gazprom now has both oil-linked and hub contracts but long-term contracts will also be replaced with spot contracts as they expire. Globally, 25% of the volumes were oil-linked including CNPCC, Qatar and PNG's contracts but Thailand, US and PGNiG contracts were linked to the Henry Hub. In China, there will be more flexibility in destination clauses (in country at least) as 7–8 new ports will be opened. In Japan, the Japan Fair Trade Commission recently carried out an investigation and decided that destination clauses breach Japanese anti-competition law particularly in Free on Board contracts. As a result, LNG sellers should adopt or sign new contracts that in compliance, they should revise their business models to include contractual clauses that provide destination flexibility. This means that from October 2018 to now, there is abundance of supply looking for a home and winter demand in Asia is generally low. This also puts pressure on Russia and its share of LNG in the EU. Novatek and the liberalisation of LNG will let Russia open up to Asia and North East Asia but most of it may still be EU bound, as LNG from Novatek continues to compete with pipeline gas. In the EU the share of LNG will also increase in the long-term outlook, and in a global LNG market Russia still can win at competition without increasing its gas share via price increases.

Perhaps a more important and relevant clarification for this book is that the gas market in the EU is not fully completed. This is central to many of the arguments that will be raised here. The internal gas market is not completed in particular in South East Europe because of a lack of liquidity. The gas market is not completed because investments are not decided by the market. In the EU all major gas infrastructure projects require some form of support, including EUGAL, the Baltic Pipeline and Polish LNG. What is left is now to fully implement the Network Codes. Instead, the rules are subject to continuous change, and improvements are welcome, but it all comes with a cost on investment, therefore making it really important for policy makers to adopt a forward-looking approach when deciding which investments to incentivise.

A final reality check: are gas and renewables friends or not? While it is true that renewables and storage compete with gas, and energy transition brings uncertainties for the gas sector with an impact on investment, renewables and natural gas are not necessarily enemies. For the decarbonisation of the electricity sector, the two commodities compete less, and for the decarbonisation of heat, transport and the industry, renewables do not offer a silver bullet, neither does green hydrogen from renewable sources as there is no capacity in the EU to produce enough green hydrogen to replace the gas in the energy mix, except in coastal countries like Denmark.

All in all, gas demand is set to stay the same until after 2030 and decrease as a result of climate policies towards 2050. To demonstrate that in numbers and case studies, it is necessary now to provide a background to consumption, production and storage levels in the EU.

3.1.2 Background to Natural Gas Consumption in the EU

Despite the fact that today's EU is the biggest natural gas importing market in the world, the increased use of natural gas in the energy mix in this region is fairly recent compared to other energy sources such as oil, nuclear and coal. Gas production in Europe began in 1959 with the discovery of the Groningen field in the Netherlands, followed, a few years later, by the first discoveries in the UK sector of the North Sea.

Natural gas was initially found as a by-product of oil as a gas deposit or as associated gas. The latter is still largely flared by oil producers around the world. It took longer for natural gas development to become economically attractive, since marketing of natural gas requires construction of expensive gas transportation and distribution systems. Transportation and storage of gas is more expensive than oil due to the lower density of gas. Gas marketing requires significant upfront capital investments. The first natural gas contracts between European gas producers and the then vertically integrated national and private energy companies were concluded on a long-term basis to allow the sunk cost to be recovered. As gas was initially considered a substitute for oil or coal, pricing of natural gas was linked to the price of these commodities.

From the 1970s EU indigenous production covered most of the region's gas demand. The use of natural gas for heating, power and manufacturing has increased to a level where around 50% of the EU's natural gas is now being imported from the non-EU countries (Norway, Russia and Algeria). Currently, gas is the dominant source of energy in most of the EU countries but production is falling rapidly everywhere including Norway. Against this, gas is still considered an important resource in the European energy mix alongside renewable energy resources. Natural gas itself can be renewable via green hydrogen.

As explained below, the demand for gas in the EU has mixed in the past due to poor economic performance until 2014, the proliferation of lower-carbon alternatives, ETS and commitment to the 2015 Paris Climate Agreement. However, decreasing indigenous production, nuclear and coal phase out is driving further imports. The need for more imports will be met by a mix of LNG and pipeline gas. Increasing Russian pipeline gas imports, despite being the cheapest option, would mean enhanced reliance to one major supplier. Other feasible alternatives for pipeline gas imports are form Azerbaijan and East-Med, and less so from Turkmenistan. Many of Gazprom's contracts with European buyers will terminate in 2019–2021 and if they do not renew the contracts, the need for imports from third countries may increase substantially. However, any alternatives to Russian supplies will require significant upstream investments, including transport infrastructure.

The role natural gas plays as a reliable, clean and affordable energy source in the EU's energy mix is undisputedly significant. Therefore, in the long-term, it is important to ensure a regulatory environment in which investment continues in the natural gas sector in the EU. Acknowledging this fact and in the view of market and regulatory uncertainties in the EU gas sector, the Energy Union Commissioner, Maroš Šefčovič, said in November 2015 that 2016 will "*be the year in which we will lay the foundations of a robust governance system bringing predictability and transparency, which is what investors need*".[4] This follows the Commission's Communication of May 2015, titled "Better regulation for better results - An EU agenda".[5] This Communication, aimed at enhancing transparency and scrutiny for better EU law-making, introduces a detailed guidelines and a toolbox for policy. It requires impact assessments to be carried out on initiatives expected to have significant economic, social or environmental impacts. These can be legislative proposals, non-legislative initiatives and implementing and delegated acts. However, an impact study, published 18 months after the Communication was presented, found that only half the EU's legislative proposals announced in 2016 were subject to impact assessment.[6]

Historically, the natural gas infrastructure is also seen by European associations as a vehicle to achieve the five dimensions of the Energy Union, as the transmission infrastructure is often perceived as key to ensuring a *"flexible, integrated and resilient gas system in Europe"*.[7] In fact, Brendan Devlin, an Advisor in the Directorate General for Energy of the European Commission, once noted: *"the greater the number of the pipelines the more liquid EU Gas markets will be"*.[8] In the next section, the regulatory elements of early transboundary pipelines in Europe will be examined to help understand the changes since natural gas became economically viable in Europe.

---

[4]European Commission, Press release "The Energy Union on Track to Deliver" (18 November 2015), available at: http://europa.eu/rapid/press-release_IP-15-6105_en.htm.

[5]Communication from the Commission to the European Parliament, the Council, the European Economic and Social Committee and the Committee of the Regions Better regulation for better results—An EU agenda, COM/2015/0215 final.

[6]Impact Assessment Institute, A Year and a Half of the Better Regulation Agenda: What Happened? 30.01.2017, available at: http://www.impactassessmentinstitute.org/br-18-months.

[7]Press release, GIE Welcomes Energy Security Package, 16.02.2016, available at: http://createsend.com/t/r-EADF0D349D9920552540EF23F30FEDED.

[8]Brendan Devlin, DG Energy, EC, 5 March 2015, Brussels, Martens Centre for European Studies, Refuelling Europe: The Single Energy Market and Energy Union in a post-South Stream Environment.

3.1.2.1 First Transboundary Pipelines in the EU

Traditionally, pipeline gas has been the dominant method for the cross-border transport of natural gas in Europe. In the case of transport via cross-border pipelines, specific production and specific markets are commonly interlocked. Pipelines are a fixed/immobile gas infrastructure that will remain on the host states territory for a period of around 30 years. They are extremely capital intensive, and although the operational costs are low, the pay-back period tends to be slow. In addition to the cost of pipeline construction and associated expenses, there are often also upstream costs for natural gas exploration and extraction, all of which has to be covered with a reasonable rate of profit via sales to designated consumers. Therefore, they require longer term commitment but also offer long-term energy supply security. Hence the UK, Germany, France, Italy, Hungary and Poland are among the top 20 countries in the world by length of oil and gas pipelines.

The first pipelines to bring natural gas into Central and Western Europe were developed under predominantly bilateral agreements, involving State producers of gas (Norway and the UK) and State consumers (Germany, Belgium, France). They contained, in many respects, similar provisions dealing with issues such as jurisdiction, pipeline ownership, operation, tariffs and taxation.[9] These projects were developed from the early 1970s to the late 1980s by vertically integrated national oil companies. For instance the Europipe is a 670-kilometre (420-mile) long natural gas pipeline from the North Sea to Continental Europe, commissioned in 1995 at a cost of 21.3 billion Norwegian krone which was owned by Gassled partners (Statoil, Petoro, ConocoPhillips, Eni,

---

[9]See e.g. Agreement between the Government of the United Kingdom of Great Britain and Northern Ireland and the Government of the Kingdom of Norway relating to the Transmission of Petroleum by Pipeline from the Ekofisk Field and Neighbouring Areas to the United Kingdom, 22 May 1973, UNTS 885, 63; Agreement between the Government of the Kingdom of Norway and the Government of the Kingdom of Belgium relating to the Transmission of Gas from the Norwegian Continental Shelf and from other areas by Pipeline to the Kingdom of Belgium, 14 April 1988, Stortingsproposisjon (St.prp.) No. 148, 187–188; Sergei Vinogradov and Gokce Mete, Cross-Border Oil and Gas Pipelines in International Law, German Yearbook of International Law (FOCUS: International Energy Law, 2013), 56.

Exxon Mobil, Norsea Gas, Shell, Total and DONG Energy) and operated by Gassco, a Norwegian state owned company, each section of the transit pipeline being subject to construction and operation regulations of the State traversed. The pipeline capacities were dedicated to developers.[10]

Completed in 1983, the Trans-Mediterranean Pipeline brings natural gas from Algeria to Italy (and further to Slovenia) with another transit pipeline via Tunisia and Sicily. Similarly, an international pipeline operational from the early 1990s is the Maghreb-Europe Pipeline laid to Spain and beyond to Portugal via an extension. It was jointly financed (US$2.5 billion) by Algeria's Sonatrach and Spain's Enagas, the line's main customer.[11] It was a dedicated pipeline constructed well before the entry into force of the Second or Third energy packages. As stated above, since 2003 most major pipeline projects within Europe are developed under derogations from the main regulations.

In Europe, the North Sea pipeline network crossed only one border at the edge of the territorial sea. However, as indigenous production declined, the need to import gas via pipelines that transit more than one border increased. Indeed, in the early 1990s the EU was exporting 109 bcm of natural gas from Russia, joined by Central Asian gas mainly from Turkmenistan (until 1994), via multiple large-diameter pipelines built from Siberia and crossing the territories of Ukraine and Belarus at entry points located on the border of Poland, Slovakia, Hungary, Romania, Bulgaria, (via Romania) and Moldova. Following the completion of a new transit corridor from Poland and Belarus linking Russia and Germany in 1999, the so-called Yamal-Europe Pipeline, this brought additional 32.9 bcm of gas to Europe.

Up to 400 bcm/year capacity of pipelines are connected with the EU area via entry points mainly in the north. Of this, nearly 30% of gas supplies are delivered from Russia via pipelines under longer term

---

[10]The Langeled, Cats, Seal, Sage, Pulsmar pipelines connect to the UK for consumption in the UK or transit. Europipe I/II, Norpipe, Zeepipe are pipelines connected directly to the continental EU import points in Emden and Zeebrügge.

[11]Sergei Vinogradov, Cross-Border Oil and Gas Pipelines: International Legal and Regulatory Regimes (AIPN Study Series, 2001).

commitments pursuant to intergovernmental agreements. To meet growing EU gas demand and find alternatives to Russian supplies, three further pipelines were proposed in the 2000s: Nabucco from the Caspian Sea region, South Stream from Russia to the Balkans and TAP from Azerbaijan. Only the latter project is being developed and the South Stream pipeline has been replaced by the Turk Stream, which connects Russia with Turkey under the Black Sea. Existing gas pipelines and future projects will be discussed in detail below following a review of viable alternatives to pipeline gas.

3.1.2.2 The Gas Bill of the EU

Natural gas makes up almost one third of the EU's primary energy consumption, and the EU depends for 50% of its consumption on three major gas-exporting countries: Russia, Norway and Algeria. Driven by the cold winter and supported by the gradual economic recovery, gas consumption in the EU increased further in 2016–2017.[12]

Due to the sharp decrease in commodity prices the gas import bill in the EU dropped from €95 billion in 2013 to €72 billion in 2015.[13] As global trends affect the cost of gas imports in the EU, creating an adequate economic and regulatory environment suitable for gas infrastructure development is essential. While international gas prices have been showing signs of convergence since 2014, oil-indexed and hub prices still present some variations.

Natural gas is the only commodity that is not initially priced on the basis of demand but rather by replacement value, with the US as the exception. This stems from the fact that gas projects have a specific cost structure. Production and transport of natural gas requires very large projects with high sunk capital costs and relatively low operation

[12]Quarterly Report Energy on European Gas Markets Market Observatory for Energy DG Energy, Volume 10 (Issue 2; second quarter of 2017).

[13]Report from the Commission to the European Parliament, the Council, the European Economic and Social Committee and the Committee of the Regions, Energy prices and costs in Europe {SWD (2016) 420 final}, 30.11.2016.

expenditures, thereby requiring full utilisation of capacity (maximum flow relative to design capacity) to justify the project by maximum revenues. Where the project is not operating at full capacity, the already lengthy pay-back period is prolonged at a loss to the operator.

According to the International Gas Union, hub prices have been significantly lower than oil indexed prices throughout the period 2008–2012.[14] In 2016, oil-indexation accounted for 30% of European gas consumption, and oil-indexed prices exceeded hub prices in Northwestern Europe.[15] Lower prices that are linked to spots markets might not provide the right signals for financing major natural gas transmission infrastructure crossing multiple territories. From a short-term perspective liquidity is desired for efficient competition. However, in case of investments aimed at securing long-term supply, renowned natural gas sector experts noted that lower prices might de-stimulate investment.[16]

In Gazprom's view oil-indexation is seen as a good buffer for any abuse by a dominant supplier. For instance, Gazprom's oil indexation contracts allow for a price formulation review if there is a new cheaper gas source coming to the border of the country. The liberalised markets, which came with the Third Energy Package, also mean that the cheapest source of gas, regardless of its origin, should be able to enter into the market as long as it does not impair competition or security of supply. However, this conflicts with the 'diversification' goal of the EU.

As a rule, then, specific production and specific consumer(s) are linked in transnational gas projects and, in terms of logistics, there is little flexibility in transportation via pipelines. This means the end of

---

[14]Commission Staff Working Document Energy prices and costs report Accompanying the document Communication from the Commission to the European Parliament, the Council, and the European Economic and Social Committee and the Committee of the Regions: Energy prices and costs in Europe (2014).

[15]Quarterly Report Energy on European Gas Markets Market Observatory for Energy DG Energy, Volume 10 (Issue 2; second quarter of 2017).

[16]Jonathan Stern and Howard Rogers, The Transition to Hub-Based Pricing in Continental Europe—A Response to Sergei Komlev of Gazprom Export, available at: https://www.oxfordenergy.org/wpcms/wp-content/uploads/2013/02/Hub-based-Pricing-in-Europe-A-Response-to-Sergei-Komlev-of-Gazprom-Export.pdf.

a pipeline is the end of a pipeline. Volumes cannot be easily diverted, and investment in the project, which is an immobile fixed infrastructure, cannot be relocated. Hence, in order to ensure profitability of the projects over their life-time for the upstream and mid-stream investors and to secure supply flows for the end users, contractual relations in European markets have typically been based on long-term commitments (around 15 to 25 years) with 'take-or-pay' clauses, although spot-based natural gas trade has co-existed in Europe (mainly the UK) for the last ten years.

Natural gas is different from any other commodity, including other fossil fuels such as coal and petroleum, which it initially intended to substitute. Indeed, when exploitation of the Groningen field began, natural gas was planned to replace coal and oil due to its promise on lower carbon emissions and higher efficiency. This is why its price was initially formulated by fuel substitution at the Groningen hub and has remained oil-linked for the past 40 years across the globe, except in the US, where the fuel substitution approach has never been applied.

EU markets are gradually moving from oil-linked price to hub-based price which is lower than the price of natural gas exported under the long-term supply agreements. Spot prices in Europe were established in the UK with the creation of the National Balancing Point (NBP) Virtual Trading Point in 1996. In the US, the Henry Hub has been in operation since the 1950s; however, the offering of standardised natural gas contracts with delivery at the Henry Hub began in 1990. There had been several attempts to establish natural gas hubs in Continental Europe until the successful launch of the Title Transfer Facility (TTF) in the Netherlands in 2003. In 2018, there were over 10 gas trading hubs across the EU. However, NBP and TTF still holds the highest volume and percentage of trading.[17]

It took around 15 years to establish the European natural gas hubs and spot prices. By 2013, around 50% of the gas traded in the market was based on spot prices. As of 2018, day ahead prices are

---

[17]Patrick Heather and Beatrice Petrovich, European Traded Gas Hubs: An Updated Analysis on Liquidity, Maturity and Barriers to Market Integration, OIES, 2017.

predominantly determined by market fundamentals. Further, the trading hubs seems to be largely integrated amongst neighbouring markets as they present almost identical price patterns. There are still a number of oil-linked contracts in the market which allows traders to arbitrage between spot and contract gas insofar as this is possible under their take-or-pay obligations.

There are a high number of active traders in the gas hubs in the EU, however the number of producers remains limited. This has an impact on the formation of spot prices as hub prices are impacted by hub liquidity. Norway, the UK and the Netherlands are already selling most of their gas at hub prices. LNG (primarily from Qatar) is sold under both sport and oil-linked contracts. In contrast, Russia has been insisting on selling its natural gas based on oil-linked prices. However, Gazprom's customers have been demanding renegotiations of Gazprom's historical oil-linked pricing terms and methodology to reflect the market-based pricing that is now prevalent in Europe. Most of these cases which were referred to arbitration have been settled without an award. Nonetheless, it is possible to predict the outcome of these negotiations, as now one third of Gazprom's contracts are hub-price linked and another third are hybrid contracts, which essentially offer a price lower than that of oil or hub-linked prices.[18] This ratio is expected to increase when most of Russia's gas contracts with European companies are renewed (if not terminated) in 2019–2021. Other than Russia, Algeria has also been traditionally selling natural gas to the EU on oil-linked contracts but this is also changing.[19]

The other key element of spot price formation is storage capacity. Storage capacity allows movement of gas between seasons—storing lower priced gas in summer periods for winter, as well as for shorter movements via fast cycle storage. Despite robust progress in the evolution of gas trading hubs in the EU there are still some problems particularly in relation to *"liquidity and data transparency; physical connectivity;*

---

[18]James Henderson and Jack Sharples, Gazprom in Europe—Two "Anni Mirabiles", But Can It Continue? OIES, 2018.

[19]Quarterly Report Energy on European Gas Markets Market Observatory for Energy DG Energy, Volume 10 (Issue 4; fourth quarter of 2017).

political willingness and cultural attitudes".[20] In order to have better price convergence across neighbouring hubs, there has to be no restrictions to physical capacity for natural gas transport between them. The only price difference between different hubs should be a reflection of the cost of transportation as per distance.

It should be noted here that wholesale prices of gas and traded gas needs to be treated with caution. In liberalised and traded markets the wholesale prices would typically be the hub price. If natural gas is imported, the border price is often used as a proxy. However, gas supplied from domestic production would be approximated to the wellhead or the entry–exit points.[21] Globally, there is a variation of benchmark hub prices. Since 2014, US Henry Hub spot prices has been lower than the UK NBP and the Netherlands TTF index (see below). Global price variations have an impact on supply, particularly on LNG flows worldwide which follows the highest price. Spot prices in the EU gas hubs have been traditionally lower than those under long-term supply contracts but the higher price translates to premium for supply security and returns on investments.

3.1.3 Gas Production in the EU

3.1.3.1 Conventional Gas

Norway, UK and the Netherlands contribute about 80% of indigenous gas production in the EU, which equals approximately 50% of demand within EU in 2014 and only 30% in 2016. European gas production has declined by more than 40% over the last decade.[22] The share of indigenous production in Europe's supply is expected to fall further due

---

[20]Patrick Heather and Beatrice Petrovich, European Traded Gas Hubs: An Updated Analysis on Liquidity, Maturity and Barriers to Market Integration, OIES, 2017.

[21]Quarterly Report Energy on European Gas Markets Market Observatory for Energy DG Energy Volume 6 (Issues 3 & 4; third and fourth quarter of 2013) and Volume 7 (Issues 1 & 2; first and second quarter of 2014).

[22]Eurostat statistic data on energy production and imports, extracted in June 2017, available at: http://ec.europa.eu/eurostat/statistics-explained/index.php/Energy_production_and_imports.

to expected price increases of indigenous gas towards 2030 as a result of the costs of replacing maturing fields, and the collapse of gas production in the Netherlands as of 2018. By 2020 gas production volume will decrease to 50 bcm. According to BP Energy Outlook 2017 a further 3.2% decrease in production per year is inevitable.[23] The UK upstream sector, in particular, is not expected to receive further investment in view of volatile oil and gas prices.[24] As mentioned above, the production output of the Groningen field in the Netherlands has also dwindled due to continuous earthquakes and seismic tremors. On top of this, the Norwegian Sovereign Wealth Fund have decided to stop investing in fossil fuels. The IEA outlook suggests that Norway will continue to reduce its gas production at a rate of 1.1% annually, corresponding to a 30% decline by 2040.[25] The drop in domestic production has resulted in attention being shifted to the viability of unconventional gas production. BP Regional Insights predicts that the decline in natural gas production in the EU leads to a deterioration self-sufficiency, with the import ratio rising from 75 to 88% by 2040.[26]

3.1.3.2 Unconventional Gas

Unconventional gas includes shale gas, which is mostly composed of methane and is extracted from intact coal beds, and gas produced when coal is gasified underground.[27] While shale gas is produced using technology developed in the 1980s, it has become economically viable

---

[23]BP Energy Outlook 2017 Edition, available at: https://www.bp.com/content/dam/bp/pdf/energy-economics/energy-outlook-2017/bp-energy-outlook-2017.pdf.

[24]Jonathan Stern, *The Future of Gas in Decarbonizing European Energy Markets: The Need for a New Approach*, 2017, available at: https://www.oxfordenergy.org/wpcms/wp-content/uploads/2017/01/The-Future-of-Gas-in-Decarbonising-European-Energy-Markets-the-need-for-a-new-approach-NG-116.pdf.

[25]World Energy Outlook, 2017.

[26]BP Regional Insights—EU, available at: https://www.bp.com/en/global/corporate/energy-economics/energy-outlook/country-and-regional-insights/european-union-insights.html.

[27]UK Houses of Parliament, Future of Natural Gas in the UK, Number 513 November 2015, available at: http://researchbriefings.files.parliament.uk/documents/POST-PN-0513/POST-PN-0513.pdf.

over the last decade thanks to a combination of horizontal drilling and hydraulic fracturing techniques.[28] The production of natural gas from shale formations revitalised the US. As conventional natural gas production is declining, the economic viability of shale gas production in Europe has attracted increased attention.

Currently, the exploration and exploitation of shale gas in Europe are almost non-existent. Recoverable shale gas reserves are estimated to be 16 trillion cubic meters (tcm) but drilling operations in the EU are very limited. Most shale gas exploration took place in Poland but with disappointing results. There is very little commercial production of tight gas and coal bed methane in the EU: around 3 tcm and 2 tcm, respectively.[29] Difficult geology has resulted in low test outputs restricting hydraulic fracturing of horizontal wells and low oil and gas prices have had negative impact on investments. More importantly, shale gas development in Europe, especially fracking, received significant public opposition on environmental grounds particularly due to concerns over groundwater pollution and atmospheric emissions. As a result, some Member States, such as France and Bulgaria have banned hydraulic fracturing.

Despite an initial onshore exploration activities for unconventional gas in the UK, who reportedly has significant resources and economically viable production. That said extractions costs, investment and infrastructure development levels and market prices remain uncertain.[30] In fact, there is a no scientific consensus on how much shale gas can be recovered across the UK. With the impact of fast-declining natural gas production in the North Sea, the central government, though not the Scottish Government, has shown support for fracking. But any production seemed unlikely before the 2020s. In addition, the UK Government lost a high court ruling over its guidance on fracking which advised controversial gas extraction technique can help combat climate

---

[28]British Geological Society data on shale gas, available at: http://www.bgs.ac.uk/research/energy/shaleGas/home.html.

[29]Technically Recoverable Shale Oil and Shale Gas Resources: An Assessment of 137 Shale Formations in 41 Countries outside the United States, June 2013, available at: https://www.eia.gov/analysis/studies/worldshalegas/pdf/overview.pdf.

[30]UK Houses of Parliament, Future of Natural Gas in the UK, Number 513 November 2015, available at: http://researchbriefings.files.parliament.uk/documents/POST-PN-0513/POST-PN-0513.pdf.

change in March 2019.[31] On the basis of a new report published by the UK's Oil and Gas Authority, which concluded that it is not possible to predict the magnitude of earthquakes fracking might trigger, the Government banned fracking in England on 2 November 2019. Fracking had little support among the broader public in the UK and it is highly questionable whether shale gas production in the rest of the EU will contribute noticeably to the EU's energy mix in the long term.

3.1.4 Natural Gas Supply Security

In 2016, natural gas held a 22% share of the European energy mix[32] and is expected to rise to as much as 30% by 2030.[33] The role of natural gas in European energy security is remarkable; there has been little change since Winston Churchill famously responded to a question on national energy security that *"safety and certainty of oil lie in variety and variety alone"*.[34] Today, however, the main concern is not about oil supplies, which have become global since World War I, but natural gas which remains regional and complex due to the geopolitics involved. The diversity of energy supplies and suppliers have been at the centre of the EU's energy policy framed under a number of documents including:

- EU Energy 2020 strategy (2010)[35]
- EU climate diplomacy for 2015 and beyond (2011)[36]
- Energy Roadmap 2050 (2011)[37]

---

[31]UK Government press release, Government ends support for fracking, 02 November 2019, available at: https://www.gov.uk/government/news/government-ends-support-for-fracking.

[32]BP Energy Outlook 2017 Edition, available at: https://www.bp.com/content/dam/bp/pdf/energy-economics/energy-outlook-2017/bp-energy-outlook-2017.pdf.

[33]EUROGAS, Long-Term Outlook for Gas to 2035, available at: http://www.eurogas.org/uploads/media/Eurogas_Brochure_Long-Term_Outlook_for_gas_to_2035.pdf.

[34]Cross-reference from, Daniel Yergin, Ensuring Energy Security, Foreign Affairs, Volume 85, Issue 2, 2006, available at: http://www.un.org/ga/61/second/daniel_yergin_energysecurity.pdf.

[35]Communication from the EC, Energy infrastructure priorities for 2020 and beyond—A Blueprint for an integrated European energy network {COM (2010) 677 final of 17 November 2010}.

[36]The EU External Action, EU Climate Diplomacy for 2015 and Beyond, Reflection Paper, available at: https://ec.europa.eu/clima/sites/clima/files/international/negotiations/docs/eeas_26062013_en.pdf.

[37]Energy roadmap 2050 {COM (2011) 885 final of 15 December 2011}.

- Climate and energy policy from 2020 to 2030 (2014)[38]
- European Energy Security Strategy (2014)[39]
- Energy Union Package (2015)[40]
- Clean Energy Package (2016)[41]

Natural gas security strategy is regulated under the Security of Gas Supply Regulation (SoS Regulation) (2010),[42] which created measures to identify threats to gas security such as operation failure of a production facility or pipelines that supply the EU. There are also a number of practices that Member States must comply with in cases of disruption. For instance, Member States need to stock at least 30 days' worth of gas for private households and other vulnerable consumers such as hospitals. Member States are obliged to prepare Emergency Plans to deal with a crisis and set up a Gas Coordination Group between national authorities and industry. An update was proposed by the Commission on the SoS Regulation in 2016.[43] In response, the Council and the European Parliament reached an agreement in April 2017.[44] The new rules aimed

---

[38]Communication from the Commission to the European Parliament, the Council, the European Economic and Social Committee and the Committee of the Regions, a policy framework for climate and energy in the period from 2020 to 2030 {SWD (2014) 15 final}, 22.1.2014.

[39]Commission Staff Working Document, In-depth study of European Energy Security, Accompanying the document, Communication from the Commission to the Council and the European Parliament: European energy security strategy {COM (2014) 330 final}, 2.7.2014.

[40]Communication from the Commission to the European Parliament, the Council, the European Economic and Social Committee and the Committee of the Regions, a Framework Strategy for a Resilient Energy Union with a Forward-Looking Climate Change Policy, Brussels {COM (2015) 80 final}, 25.2.2015.

[41]European Commission—Press release, Clean Energy for All Europeans—Unlocking Europe's Growth Potential Brussels, 30 November 2016, available at: http://europa.eu/rapid/press-release_IP-16-4009_en.htm.

[42]Regulation (EU) No. 994/2010 of the European Parliament and of the Council of 20 October 2010 concerning measures to safeguard security of gas supply and repealing Council Directive 2004/67/EC Text with EEA relevance.

[43]Proposal for a REGULATION OF THE EUROPEAN PARLIAMENT AND OF THE COUNCIL concerning measures to safeguard the security of gas supply and repealing Regulation (EU) No. 994/2010, COM/2016/052 final—2016/030 (COD).

[44]Regulation (EU) 2017/1938 of the European Parliament and of the Council of 25 October 2017 concerning measures to safeguard the security of gas supply and repealing Regulation (EU) No. 994/2010, OJ L280, 28.10.2017.

to bring more transparency to long-term gas supply contracts. In addition, they introduced a solidarity principle covering the EU's neighbouring states (Energy Community area) in the case of a gas crisis and closer regional cooperation via the development of joint preventive and emergency measures.

According to ACER, ensuring gas security can best be achieved by having a more diverse upstream supply in Europe so that over-dependence on one source of supply and the risks of disruption due to physical restrictions or political interference can be avoided.[45] This requires new investments in gas infrastructure for connecting regions and such investments require a long term vision for gas demand and supply. Short term focus makes the EU more dependent on a single source of imports.[46] Indeed, "*a well-connected EU energy market, where energy flows freely across borders and no Member State remains isolated from the EU energy networks, is a pre-condition for creating a resilient Energy Union policy*".[47] However, the current energy infrastructure in Europe is not yet designed for fully integrated markets.[48] Substantial funds are needed to construct the missing gas infrastructure links.

The lack of intra-EU infrastructure connections is not the main problem as this is a matter of budget which is allocated via CEF and TEN-E frameworks. The principal challenge to liquidity of energy markets is the EU's import dependence which will likely grow with decreasing domestic production. The diversification of gas supplies and access to new sources is of paramount importance for EU energy security in the long term. The diversification policy of the EU is traditionally focused on developing alternative routes such as the Southern Corridor and

---

[45]Press release, ACER launches updated Gas Target Model, Ljubljana, 08.01.2015, available at: http://www.acer.europa.eu/Media/Press%20releases/ACER%20PR-02-15.pdf.

[46]Ten year network development Plan 2015, Entso-G available via, http://www.entsog.eu/publications/tyndp.

[47]European Commission—Press release, Energy: Central Eastern and South Eastern European countries join forces to create an integrated gas market, 10.07.2015, Dubrovnik, available at: http://europa.eu/rapid/press-release_IP-15-5343_en.htm.

[48]Special Report No. 16/2015: Improving the security of energy supply by developing the internal energy market: more efforts needed, available at: http://www.eca.europa.eu/Lists/ECADocuments/SR15_16/SR_ENERGY_SECURITY-EN.pdf/.

the construction of new LNG import facilities. Measures also include developing liquid gas hubs in the Mediterranean and demand side measures like decreasing energy consumption via energy efficiency and increased share of renewable energy in electricity production as opposed to imported gas. However, it is now shifting to diversifying the gas portfolio with renewable and low carbon gases, such as biomethane, hydrogen and synthetic gases and the 2020 gas package is expected to provide the framework for these new types of gases to integrate the gas grid and transmission networks and localise the gas sector.

3.1.4.1 Storage

Gas storage also plays a significant role in EU energy security, helping to mitigate interruptions and balance seasonal demand variations.[49] The role of gas as seasonal storage for renewable energy is essential as the electricity system requires a constant balancing of generation and demand and storage options for electricity on the network are very limited. One such option is pumped hydroelectric storage capacity; however its capacity is not substantial in the EU. For battery storage, the costs are still very high and it is not expected to be competitive at least until 2040, however the costs are expected to fall by 50% by 2030. If this happens earlier, this could be a major game changer for natural gas globally.[50]

Natural gas can be stored in liquid form in above-ground tanks or in underground facilities. Depending on the geology, depleted oil and gas reservoirs (onshore and offshore), aquifers or salt cavern formations, leached out in salt domes, could be used. Depleted fields already have

---

[49]Communication from the Commission to the European Parliament, the Council, the European Economic and Social Committee and the Committee of the Regions on an EU strategy for liquefied natural gas and gas storage {SWD (2016) 23 final}.

[50]International Renewable Energy Agency (IRENA), Electricity storage and renewables: Costs and markets to 2030, October 2017, available via: https://www.irena.org/publications/2017/Oct/ Electricity-storage-and-renewables-costs-and-markets; see also IEA, Commentary: Battery Storage Is (Almost) Ready to Play the Flexibility Game, 07.02.2019, available at: https://www.iea.org/ newsroom/news/2019/february/battery-storage-is-almost-ready-to-play-the-flexibility-game.html.

a number of useable wells, field gathering facilities, and pipeline connections which reduces the cost of conversion to gas storage. The geology is well known. These fields have previously trapped hydrocarbons which minimises the risk of reservoir leaks. Although the requirements vary, typically these reservoir types require 50% cushion gas (i.e. equal amounts of base gas and working gas). Due to the nature of the reservoir producing mechanisms, working gas volumes are usually cycled only once per season (with the exception of storage reservoirs with extremely high withdrawal rates). These reservoirs are often old and require a substantial amount of well maintenance and monitoring to ensure working gas is not being lost via well-bore leaks into other permeable reservoirs. An aquifer is suitable for gas storage if the water bearing sedimentary rock formation is overlaid with an impermeable cap rock. The gas is injected at the top of the water formation and displaces the water down structure. While the geology of aquifers is similar to depleted production fields, their use in gas storage usually requires more cushion gas and greater monitoring of withdrawal and injection performance. They are typically close to end user market or existing pipeline infrastructure. They have high deliverability from the combination of high-quality reservoirs, and active water drive during the withdrawal cycle. The high deliverability increases the ability to cycle the working gas volumes more than once per season. Salt caverns are solution-mined cavities in existing salt domes and structures. These shallow cavities are filled with injected natural gas and act as high pressure storage vessels for natural gas, providing very high withdrawal and injection rates relative to their working gas capacity. They also have correspondingly low cushion gas requirements of 25%, which can approach zero in emergencies and ultra-high deliverability (much higher than depleted reservoir and aquifer storage). These reservoirs have the highest operational flexibility as they can cycle working gas four to five times a year. Salt caverns provide excellent seals as their walls are essentially impermeable barriers, and, therefore, the risk of reservoir gas leaks is small. Additionally, the gas network itself can store substantial amounts of linepack.

The most commonly used underground gas storage sites are depleted oil and gas reservoirs that are close to consumption centres or existing pipeline infrastructures, which helps in avoiding extra gas transportation costs to bring the gas from storage to market.[51] Pursuant to Article 2(9) of the Gas Directive states:

storage facility' means a facility used for the stocking of natural gas and owned and/or operated by a natural gas undertaking, including the part of LNG facilities used for storage but excluding the portion used for production operations, and excluding facilities reserved exclusively for transmission system operators in carrying out their functions.

According to data from Gas Storage Europe (GSE), the total working gas storage capacity in 2017 in the EU was about 112 bcm. One third of this capacity is located within three Member States: France, Germany and Italy.[52]

From a regulatory perspective, there are three types of gas storage facilities in the EU: commercial, strategic (national strategic stocks) and excluded (TSO or production storage). Storage has a crucial role to play both in facilitating development of competition and in contributing to security of supply. Storage facilities provide flexibility, not only from a security of supply and cost perspective but because of the flexibility it brings to natural gas which plays an important balancing role for intermittent renewables in electricity production. However, many gas storage facilities have been forced to close down due to market failures as commercial storage prices do not reflect the full cost of gas storage.[53] This is largely because the tariffs network codes, which define how charges should be calculated at entry-exit points for storage sites across the EU, are insufficient to harmonise tariffs for putting in and taking out gas from storage sites across the EU. The shippers find the tariffs according to storage industry, and this acts as a major deterrent for investment in

[51]https://www.eia.gov/naturalgas/storage/basics/.

[52]Interactive Gas Storage Map, available at: http://www.gie.eu/index.php/maps-data/gse-storage-map.

[53]Gas Infrastructure Europe, Gas Storage Market Failures, Report Summary, 2017, available at: https://www.gie.eu/index.php/publications/gse/doc_download/26901-gie-poyry-study-executive-summary-on-gas-storage-market-failures.

storage facilities.[54] Strategic storage is also not considered cost effective. Whilst it is beyond the scope of this book to explore market failures in gas storage in Europe, it is important to highlight the role gas storage plays in gas supply security and to demonstrate how the regulatory framework does not always reflect the commercial reality of European gas markets.

3.1.4.2 Import Options

As a result of a sharp 3.2% annual decline in domestic production, the share of imported gas into the EU is expected to rise to 80% from the current 50% level.[55] In light of this, several questions need to be considered: What are the import options for Europe to meet its likely import deficiency towards 2030? What do global increases in the supply of LNG mean for Europe? What has been the impact of pipeline politics in the last decade on decisions to invest in further natural gas import pipeline projects? How has the relationship between producers, consumers and transit states within Europe changed, and what does it mean for future import options for the EU? Can the new supply routes from East-Mediterranean, Caspian region and Middle East provide alternatives to Russian gas? And finally, what role for renewable and other low carbon gases?

LNG

With global LNG production rapidly increasing, it could play an important role in the EU gas balance due to the flexibility it provides. In 2016, there was a 6% increase in global LNG trade. The amount of new liquefaction capacity going online reached as much as

---

[54]Gas Infrastructure Europe, Gas Storage Market Failures, Report Summary, 2017, available at: https://www.gie.eu/index.php/publications/gse/doc_download/26901-gie-poyry-study-executive-summary-on-gas-storage-market-failures.

[55]BP Energy Outlook 2017 Edition, available at: https://www.bp.com/content/dam/bp/pdf/energy-economics/energy-outlook-2017/bp-energy-outlook-2017.pdf.

21 bcm in 2016.[56] Despite this, some forecasts suggest that there will be no significant LNG imports into the EU post-2025.[57] The figures for 2013–2014 saw very high global LNG prices as opposed to those for pipeline gas caused by a surge in gas demand in Asian markets. The difficulty in attracting spot LNG cargoes is that they follow demand. Unlike regional pipeline gas, spot LNG is truly global and follows the markets. Hence, the price impact of LNG has to be considered in any future forecast.

The price difference between LNG and pipeline gas, and the fact that LNG supplies are tied under long-term contracts meant underutilisation of regasification capacity in Europe. In 2014, only 32% of LNG terminals were used, whereas pipeline capacity was almost fully utilised.[58] Reliance on single suppliers is seen as a major risk to security of supply according to the 2014 'Energy Security Strategy'.[59] Since the 2014 transit crisis between Russia and Ukraine, interest in LNG has increased to promote diverse sources of supplies, especially in Eastern European countries dependent on Russia for their imports. As a result, there has been substantial progress in increasing LNG capacity in Lithuania, Poland and Finland. The hope is that with the US shale gas boom, US LNG cargoes can be shipped to Europe. The US LNG export projects could see more than 100 bcm of exports by 2030. However, as explained above, LNG is 'price sensitive', and, therefore, if prices are higher in Asia, LNG cargoes will be directed to these markets. It is also important to take into account the possible reaction of Russia who can bring future gas prices to levels that can compete with US LNG (US LNG is already no longer competitive with Russian gas today). Therefore, the prospect of 'cheap' US LNG imports to Europe

---

[56]SNAM and BCG Center for Energy Impact, Global Gas Report 2017, available at: http://www.snam.it/export/sites/snam-rp/repository/file/gas_naturale/global-gas-report/global_gas_report_2017.pdf.

[57]Global gas and LNG themes—Q1 2017, Wood Mac.

[58]Opinion of the Committee on the Environment, Public Health and Food Safety (8.9.2016) for the Committee on Industry, Research and Energy on the EU strategy for liquefied natural gas and gas storage (2016/2059[INI]).

[59]Communication from the Commission to the European Parliament and the Council European Energy Security Strategy {COM (2014) 330 final}.

is questionable, especially bearing in mind the high shipping costs to Europe and given the impact of sustained low oil prices on planned capacity costs.

In a statement released on July 2018, the EU Commission President Jean-Claude Juncker announced the intention of the EU to construct more LNG import terminals to boost US LNG exports to Europe.[60] This is a political statement without any legal implications, as neither the EU nor the Commission has the right to determine Member States' conditions for exploiting its energy resources, its choice between different energy sources and the general structure of its energy supply under Article 194 (2) of the TFEU. Furthermore, there is already ample LNG regasification in the EU which remains underutilised. Ultimately, it will be the price that will determine how much US LNG will be exported to the EU.

The US has ambitious plans to supply gas to the EU, but there are new players emerging too. Australia, Canada and Mozambique are all increasing LNG production. New LNG is needed but cost efficiency and flexible business models are essential. Furthermore, the EU will not automatically benefit from this rise, as these countries cannot be forced to supply to Europe. LNG shippers will choose the most profitable markets. Currently high prices in Asia are driving the LNG cargoes at a rate 30% more expensive than European spot markets. If Asian gas demand remains high, which is likely, additional import capacity will have to be constructed to substitute for the decreasing production in Europe. Even if LNG imports to Europe increase, pipeline gas will play an important role due to lack of adequate LNG regasification infrastructure and difficulties in moving the gas from the west of Europe to the east.[61] The next sections will investigate pipeline gas import options.

---

[60]News Article, CNN, Trump and Juncker Agree to Take Steps to Boost US LNG Exports to Europe, 25.07.2018, available at: https://www.cnbc.com/2018/07/25/europe-will-import-more-us-natural-gas-trump-and-juncker-say.html?__source=sharebar|twitter&par=sharebar.

[61]Beatrice Petrovich and Howard Rogers, Harald Hecking, Simon Schulte and Florian Weiser, Future European Gas Transmission Bottlenecks in Differing Supply and Demand Scenarios (EWI & OIES, 2017), 26.

Pipeline Gas

Pipeline gas currently makes up the bulk of EU gas imports. In 2015, around 87% of the EU's gas was imported via pipelines with only 13% from LNG.[62] Total pipeline trade in Europe reached over 415 bcm. As mentioned above, there are three main suppliers: Russia, Norway and Algeria. In 2016, Russian exports to Europe and Turkey was around 185 bcm compared to 115 bcm from Norway[63] Forecasts indicate at least a 10% decrease in Norwegian supplies by 2035, however until this time Norway is committed to maintaining current production levels. The following sub-sections will review options for pipeline gas supplies from non-EU producers, including Russia, Turkmenistan, Iraq (Kurdistan), Iran, Azerbaijan and the East Mediterranean area, including Cyprus, Israel and Egypt.

Russia

Russian gas exports to the EU have held an important share in the EU energy mix since the 1990s. It is reported that from this time, volumes have never fallen below 100 bcm each year (however, all European long-term gas contracts are subject to commercial confidentiality, which means that it is difficult to be definitive about their terms). Currently Russia exports around 135 bcm of gas to the EU per annum.[64] In 2017, Russia remained EU's largest supplier with a share of 43% of total imports.[65] Most Russian gas is shipped under long-term contracts, with

---

[62]Quarterly Report Energy on European Gas Markets Market Observatory for Energy DG Energy Volume 9 (Issue 1; fourth quarter of 2015 and first quarter of 2016).

[63]BP Statistical Review of World Energy, June 2017, available at: https://www.bp.com/content/dam/bp/en/corporate/pdf/energy-economics/statistical-review-2017/bp-statistical-review-of-world-energy-2017-full-report.pdf.

[64]Luca Franza, Outlook for Russian Pipeline Gas Imports into the EU to 2025, Clingendael International Energy Programme, 2016, available at: http://www.clingendaelenergy.com/inc/upload/files/CIEP_paper_2016_2B_Russia_web.pdf.

[65]Quarterly Report Energy on European Gas Markets Market Observatory for Energy DG Energy, Volume 10 (Issue 2; second quarter of 2017).

Ukraine being the principle supply route.[66] In 2017/2018, around 15% of Russian gas is expected to be sold via auctions, but this share may change significantly by 2035 when Russia's longer-term gas supply contracts with EU countries will expire.

While the existence of long-term supply contacts means that Russia is likely to protect its 30% market share towards 2035, the EU would like to diversify away from Russian gas as part of its energy diversification policy goal under the Security of Supply Directive. Russia, in turn, would like to diversify away from the Ukrainian transit route and has made it very clear that it does not have any intention to expand its transit contract with Ukraine after 2019. This is the philosophy behind the North Stream, South Stream (now defunct) and Turk Stream pipelines which will not bring new supplies but help to reduce transit risks and costs through establishing new routes. However, other than the North Stream 2 pipeline, Russia is unlikely to build another pipeline to Europe due to uncertainties in demand, although the extension of the Turk Stream pipeline to accommodate volumes for South East Europe will remain as its key objective. Even if North Stream 2 is built, Russian supplies to EU could still be limited due to infrastructure limitations.

The EU Commission currently opposes the construction of the North Stream 2 pipeline as it does not serve their strategy for diversification of supply and changing the gas balance in Central and Eastern Europe.[67] However, it has limited ability to block the project as the North Stream 2 is an offshore pipeline and therefore will be outside the regulatory scope of the Third Energy Package. Gazprom have reached an agreement with E.ON, Shell, OMV and BASF/Wintershall to add a second line to the existing North Stream pipeline, hence increasing its transit capacity by an additional 55 bcm from 2020. While the pipeline falls outside the regulatory scope of the Third Energy Package (described as 'a legal void' by the Commission), the Commission

---

[66]Quarterly Report Energy on European Gas Markets Market Observatory for Energy DG Energy, Volume 10 (Issue 2; second quarter of 2017).

[67]News Article, Politico, Šefčovič Warns Energy Firms Over Nord Stream II Participation, 09.07.2015, available at: https://www.politico.eu/article/sefcovic-warns-energy-firms-over-nord-stream-ii-participation/.

insisted that it must comply with the objectives of the Energy Union.[68] The estimated €9.5 billion project is supported by the governments of Germany and Austria. While the Commission asked for a mandate to negotiate the terms of the project, there was little scope for the Commission to receive such authority as the project is classified as a commercial endeavour.[69] There is no IGA between governments of Russia and Germany. The project is built as per joint venture agreements between Gazprom and other West European companies.

The South Stream pipeline project, which was designed to reduce Russia's reliance on Ukraine, was cancelled as it did not meet EU requirements. The Turk Stream pipeline emerged as a substitute in 2015 with a 15.7 bcm or 31.5 bcm capacity depending on whether it consists of one or two lines, with the first line to be completed by 2019.[70] There is fear now from the Commission that access to EU markets could rest with Turkish authorities, as Russia plans to sell its gas via Turkey at the Greek border to the EU. Whether the second line of the Turk Stream goes ahead or not, Russia is likely to remain a major integral player in the European gas markets at least until 2035.

Middle East

The Middle East has the largest total proven natural gas reserves by region, amounting to a total of 79.4 trillion cubic meters, more than 40% of the world's total proven reserves.[71] Iran has the largest share of these, having the world's second biggest natural gas reserves, estimated

---

[68]Press release, Remarks by President Donald Tusk After the European Council Meeting, 18 December 2015, available at: http://www.consilium.europa.eu/en/press/press-releases/2015/12/18-tusk-final-remarks-european-council/.

[69]New Article, Politico, Nord Stream 2 Fight Set to Heat Up as Countries Show Their Cards, 09.04.2017, available at: http://www.politico.eu/article/nord-stream-2-fight-set-to-heat-up-as-countries-show-their-cards/.

[70]News Article, Gary Peach, Russian Pipes Divide Europeans, June 2015, Energy Intelligence, available at: http://beta.energyintel.com/world-energy-opinion/russian-pipes-divide-europeans/.

[71]BP Statistical view of World Energy 2017, https://www.bp.com/content/dam/bp/en/corporate/pdf/energy-economics/statistical-review-2017/bp-statistical-review-of-world-energy-2017-full-report.pdf.

at 33.5 trillion cubic meters in 2016.[72] Iran also has a long-term plan to supply Europe with gas since the now defunct Nabucco project was first proposed in early 2000s. The UN sanctions imposed on Iran over its nuclear programme have put on hold these plans, but after the sanctions were partially lifted in 2016, Iran returned into the picture as an alternative pipeline gas supplier to Europe. In 2019, a completely different political picture emerged as the US reinstated sanctions on Iran and in response Tehran has said it will exceed uranium enrichment levels agreed under the 2015 Iran nuclear deal. However, in August 2018, the US granted an Iran sanctions waiver to the Southern Gas Corridor natural gas pipeline projects.[73] Iran's national oil company NICO holds a 10% stake in the Shah Deniz consortium that is developing the Shah Deniz 2 gas deposit in Azerbaijan. This suggests that when strategic energy security issues, particularly when EU economies are concerned, arise US sanctions could be circumvented, yet today tensions continue to rise rapidly in the Gulf which makes Iran's role as a future supplier of natural gas to the West unfeasible for the foreseeable future.

Iraq was also previously considered as a potential supplier in the Nabucco project. Proven gas reserves are recorded as more than 3.8 trillion bcm.[74] In recent years, Iraq has been developing its gas reserves after years of sanctions and wars. The TANAP project was seen as an opportunity to accommodate gas from Iraq's Kurdistan Regional Government (KRG) region en-route to EU gas markets. Indeed, a deal was signed in September 2013 with KRG to construct a gas pipeline by Botas and Siyahkalem, a Turkish company to receive a licence to import pipeline gas from Northern Iraq by 2017. As of 2019, the pipeline has not been constructed. However, the Turkish government set up a special

---

[72]BP Statistical view of World Energy 2017, https://www.bp.com/content/dam/bp/en/corporate/pdf/energy-economics/statistical-review-2017/bp-statistical-review-of-world-energy-2017-full-report.pdf.

[73]News Release, S&P Global, US Confirms Iran Sanctions Waiver for Azerbaijan Natural Gas Pipeline to Turkey, Europe, 07.08.2018, available at: https://www.spglobal.com/platts/en/market-insights/latest-news/natural-gas/080718-us-confirms-iran-sanctions-waiver-for-azerbaijan-natural-gas-pipeline-to-turkey-europe.

[74]Iraq facts and figures, OPEC, available at: http://www.opec.org/opec_web/en/about_us/164.htm.

purpose company, the Turkish Energy Company (TEC), with the aim of importing around 4 bcm of gas to Turkey from KRG. This gas could then be shipped to Europe via TANAP, when it is commissioned, or would potentially free-up Turkey's share in Azeri gas with the excess shipped to the EU via TAP.

While the scale of natural gas reserves in the Middle East is undisputed, there are two major constraints to be overcome before any supply can be directed to Europe. First, the reserves are underdeveloped and in need of major upstream investment. Second, the Middle East itself, particularly Iran and Iraq as potential suppliers, have a rising demand for the domestic consumption of gas. Iran is the third largest consumer of natural gas after Russia and the US and slightly above the annual consumption evidenced in China. This is due to subsidised internal prices, an increasing population, the practice of gas re-injection into oil fields and rising industrial consumption. Iraq suffers greatly from the lack of electricity generation plants. Iraqi people have only 14.6 hours of electricity per day on average, although the situation is getting better in the KRG region. In order to enhance access to electricity, the Ministry of Electricity in the KRG places emphasis on increasing installed natural gas fired power generation. Hence, any growth in production would first target domestic markets. The surplus then needs to be utilised to further improve production potential and investment in energy infrastructure. They will naturally seek to maximise their revenues from cross-border supplies to the greatest extent possible. The gas then will be directed to where prices are higher and the return of investment is higher. It goes without saying that they have alternatives to European markets looking east to Asia, to supply China, Pakistan and India, where the price is higher and consumption is greater. In addition, any deal between Turkey and the KRG on pipeline construction seems unlikely in the aftermath of the Kurdish independence referendum in October 2017.

Caspian Region

The Caspian Sea region includes Russia, Azerbaijan, Kazakhstan, Turkmenistan, Uzbekistan and Iran. However, as Russia stands alone and Iran is included within the Middle East discourse, only the remaining Caspian States will be considered in this section. As of 2016, proved natural gas reserves stood at 1.1 tcm (Azerbaijan), 1.0 tcm (Kazakhstan), 1.1 tcm (Uzbekistan) and 17.5 tcm (Turkmenistan).[75] Azerbaijan produced 17.5 bcm of gas in 2016, Turkmenistan 66.8 bcm, Uzbekistan 62.8 bcm, and Kazakhstan 19.9 bcm.[76]

Turkmenistan, Uzbekistan and Kazakhstan already export natural gas to China via the Central Asia-China Gas Pipeline. Turkmenistan alone exports more than 30 bcm per year to China, and these volumes are expected to double by 2020. Kazakhstan's KazMunaiGaz signed an MoU with China's China National Petroleum Corporation (CNPC) to expand the Kazakhstan-China Pipeline to 55 bcm per year.[77]

Azerbaijan has been exporting natural gas to Turkey and Georgia via the South Caucasus gas pipeline since 2006. The pipeline which has been expanded to accommodate volumes destined for Europe has a capacity of 25 bcm per year. However, the whole of the Southern Corridor is estimated to have the potential to bring almost 100 bcm of gas from the Caspian region to Europe, potentially allowing East-Med and Middle East gas to flow through, as TANAP is essentially a smaller version of Nabucco.

Despite Caspian States seeming to be tied up in meeting Chinese demand, diversification is also part of these producers' agenda. Dependency on a single consumer poses revenue security risk. Hence, Turkmenistan, in particular, has continued dialogue on gas exports to Europe. Currently, Russia buys gas from Uzbekistan and Turkmenistan but plans not to renew its purchase contracts upon their expiry. Turkmenistan is also engaged in the Turkmenistan-Afghanistan-Pakistan-India Pipeline project

---

[75]BP Statistical View of World Energy 2017, https://www.bp.com/content/dam/bp/en/corporate/pdf/energy-economics/statistical-review-2017/bp-statistical-review-of-world-energy-2017-full-report.pdf.

[76]BP Statistical View of World Energy 2017, https://www.bp.com/content/dam/bp/en/corporate/pdf/energy-economics/statistical-review-2017/bp-statistical-review-of-world-energy-2017-full-report.pdf.

[77]News Article, Platts, Kazakhstan to Export Up to 5 bcm/year of Natural Gas to China as of, 07.06.2017, available at: https://www.platts.com/latest-news/natural-gas/moscow/kazakhstan-to-export-up-to-5-bcmyear-of-natural-26749259.

as part of its market diversification policy. Hence, to attract supplies from the Caspian region, the EU has to offer itself as a very promising option, which is unlikely to take place considering the demand unpredictability and decreasing willingness of European companies to commit to long term bookings.

East-Mediterranean

East-Mediterranean gas also presents an opportunity for Europe as a potential pipeline gas and LNG supplier. Egypt has 1.8 tcm and Israel has 0.2 tcm of total proven gas reserves as of 2017.[78] Cyprus also has abundant gas deposits. While Egypt has the largest proven reserves in the Eastern Mediterranean region, in 2015, after being a natural gas exporter, became an importer—as production declined and domestic consumption peaked due to its electricity sector reform. In 2014, Egypt only exported 0.4 bcm of LNG to Asia Pacific. In 2016, new projects came online with production gradually increasing and Egypt plans to double gas output by 2020, yet new volumes of Egyptian gas are likely to be offered to the world markets via LNG. Israel, on the other hand, started discussions on exporting natural gas within the region and to international markets shortly after the natural gas discoveries in 2009. A possible pipeline through Turkey to bring Israeli gas to Europe has been subject to high-level negotiations. They were progressing in 2017 into talks on the price and the route of the proposed natural gas pipeline.

A potential stumbling block in the success of this project is the situation in Cyprus, since any pipeline crossing through the EEZ of Cyprus is required to obtain its consent on the delineation of the course. The Turkish route could be open to hypothetical supplies from the Caspian and Middle East, in addition to the East-Med gas. However, taking into account the rising tensions between the EU and Turkey in 2019

---

[78]BP Statistical View of World Energy 2017, https://www.bp.com/content/dam/bp/en/corporate/pdf/energy-economics/statistical-review-2017/bp-statistical-review-of-world-energy-2017-full-report.pdf.

Table 3.1 Natural gas supply options to the EU

	Appeal for supplying the EU	Cost of supplies to the EU	Reserves	Alternative markets	Geopolitical risks
Turkmenistan	Diversification of market outlets	High cost of construction of TCP pipeline	Very large	Asia and South Asia	Medium: Caspian water statues resolved
KRG	Fast monetisation of resources	Cost of transport infrastructure is uncertain	Medium	Turkey	Medium: Dispute with Bagdad and physical security of pipelines
Iran	Monetisation of resource and integration to world markets	Cost of transport infrastructure is uncertain	Very large	Many options if LNG is developed	High: US sanctions
Azerbaijan	Political and commercial motivation	Low costs of TANAP/SPC expansion	Medium to small	Turkey	Small
East-Med	Fast monetisation and regional use of resources	High cost of off-shore pipeline construction	Medium to large	Turkey and Jordan, and many via LNG	High: Cyprus issue and internal opposition to exports

Source Clingendael International Energy Programme

over Turkey's allegedly illegal current drilling activities in the Eastern Mediterranean, this option seems unlikely in the foreseeable future. The EU is even prepared to cut pre-accession funding and cancel high-level meetings with Ankara in response.

The other alternative for construction of an East-Med pipeline is a European PCI from Israel through Cyprus to the Greek island of Crete and further on to the EU. However, this is seen as complex, risky in terms of safety and environment and very expensive. Unless the EU pays for the long distance pipeline it is highly unlikely that either Israeli or EU shippers would commit to a gas pipeline in the current regulatory and market environment. The EU's economy and the new Clean Energy Package[79] would make it difficult for the EU institutions to justify putting public money into such a risky project.

Table 3.1[80] summarises various factors likely to impact future natural gas supplies to the EU from producers other than Russia.

3.1.5 Future Demand and Decarbonisation Goals

As discussed previously, natural gas can play a role important role alongside intermittent renewables: with around 50% of the GHG emissions of coal and significantly less emissions than oil, its ability to be stored, flexibly transported (for example, via pipelines and LNG), means that it can serve as a transition or bridging and grid-balancing fuel in addressing climate change. The intermittent renewable energy sector needs natural gas at least until large scale electricity storage becomes viable. To date, however, the full potential has not been realised. There are also non-intermittent renewables such as biomass, hydro, geothermal— and the less intermittent wave and tidal renewable technology options. However, in 2012, geothermal energy contributed only to 0.2% of

---

[79]Press release, Commission Proposes New Rules for Consumer Centred Clean Energy Transition, 330.10.2016, available at: https://ec.europa.eu/energy/en/news/commission-proposes-new-rules-consumer-centred-clean-energy-transition.

[80]Luca Franza, Outlook for Russian Pipeline Gas Imports into the EU to 2025, Clingendael International Energy Programme, 2016, available at: http://www.clingendaelenergy.com/inc/upload/files/CIEP_paper_2016_2B_Russia_web.pdf.

total net production in the EU.[81] There are also sustainability problems with biomass and large-scale hydro, and wave and tidal are still in their infancy regarding technological and market readiness.[82]

Although EU gas demand projections are estimated to be high (500–716 bcm), albeit that the projections are unpredictable due to a wide range of factors (see Sect. 3), in all circumstances a sharp decrease in indigenous production will mean that demand for natural gas imports will increase towards 2050 for heat, transport and power generation unless alternatives are found. The three pillars of the EU decarbonisation agenda are to be seen in conjunction with the Clean Energy Package that confirms natural gas as an important enabling fuel towards a low carbon economy.

What is clear from the latest set of EU regulation on energy is that the decarbonisation is taken seriously by the Commission and many DG Energy officials acknowledge that natural gas will play an important role in the EU energy mix towards 2050, in particular in transport, heating and electricity sectors and as a back-up for intermittent renewables. Hence, in view of declining production, to avoid deficits and increasing costs for consumers via expensive LNG options, in the short to medium term competitive markets, a predictable investment climate and an enabling infrastructure should be put in place to avoid further public funding for uncommercial security of supply projects.

At the same time, the Paris Agreement look towards limiting temperature increase by 2 °C with ambitions for a 1.5 °C target. The EU adopted ambitious Renewable Energy and Energy Efficiency Directives, which strive towards the Paris Agreement, driven by the EU Parliament. Despite all efforts, the world is actually heading towards 3 °C, which will have serious consequences on the environment and

---

[81]Eurostat, Primary Production and Gross Inland Consumption of Geothermal Energy, available via: https://ec.europa.eu/eurostat/web/environmental-data-centre-on-natural-resources-old/natural-resources/energy-resources/geothermal-energy.

[82]Geoffrey Wood, Policy Risk, Politics and Low Carbon Energy, in Jennifer I. Considine and Keun-Wook Paik (eds.), *Handbook of Energy Politics* (Edward Elgar, 2018).

human social, economic and political systems.[83] In terms of consumption, the IEA noted that in 2015 global energy consumption trends were not impacted by the Paris Agreement. In fact, in 2018 global energy demand increased by 2.3%, the fastest level witnessed this decade. This was driven by a robust global economy, weather conditions and moderate energy prices. They also noted that the Golden Age of Gas report actually got it right in predicting current gas demand and its direction of travel. In China coal to gas switching could play a meaningful role.

Despite the current upward trend in natural gas consumption, there is an overall acknowledgement that the dependence towards fossil energies should be significantly reduced. After 2030 gas will become high carbon. What can be the future role of gas in this context? What is 'renewable gas'? Hydrogen, biogas, carbon capture and storage? Does 'renewable gas' justify fossil gas infrastructure? Taking into account the EU 2050 net zero carbon ambition, the next section will look closely into alternatives to natural gas at scale.

3.1.6 Renewable and Other Low Carbon Gases

3.1.6.1 Hydrogen

A scalable alternative to natural gas is hydrogen, which is already commonly used in some industrial processes for instance in ammonia production, low-carbon methanol, other chemicals and in the iron and steel industry. There are now an overwhelming number of studies concluding that the technological solutions for replacing natural gas with renewable gases are already in place. These gases could allow the EU to continue to consume a diversified amount of gas in line with its net-zero carbon goal by 2050.

---

[83]Geoffrey Wood and Keith Baker, *The Palgrave Handbook of Managing Fossil Fuels and Energy Transitions* (Palgrave Macmillan, 2019).

Hydrogen can be produced from coal and natural gas. It is actually an artificial fuel similar to gasoline. Hydrogen that is produced with CCS process, by decarbonising natural gas via steam forming processes of natural gas, could be used in the heat sector and could be transported through conversion of the existing system. This technology is commonly called Blue Hydrogen.[84] The technology for both capturing CO_2 and adaptation of existing transmission and distribution networks is currently very expensive. Hydrogen can also be produced via electrolysis by using renewable energy. This technology is commonly referred to as Green Hydrogen.[85] Another important area where hydrogen could be useful is the transport sector as hydrogen can be used to power electric motors in cars, buses, boats, and other vehicles. Hydrogen can also be converted into electricity and heat in a fuel cell, which means that the advantages of hydrogen can also play an indirect role in the electricity and the building heating system. Hydrogen production via renewable energy, which is often referred to as renewable gas, is mainly pollution-free. In the UK, renewable gas is considered as part of a security of supply and emission reduction policy which could help to mitigate decreasing indigenous gas production and be used to meet domestic consumption in the heating sector.[86]

While in the EU it may not be possible to produce renewably produced hydrogen at the scale needed, neighbouring regions in North Africa, South Africa and Australia can do it thanks to ample sources of sun and wind. The question is how to bring it from there to the EU? It can in principle be done via gas pipelines and pipelines can be buried so that public acceptance is high. Around 350 times more gas can be stored compared to electricity in the EU currently. Green hydrogen

---

[84]IEA, Commentary: The clean hydrogen future has already begun, by Noé van Hulst Hydrogen Envoy for the Ministry of Economic Affairs & Climate Policy, The Netherlands, 23.04.2019, available at: https://www.iea.org/newsroom/news/2019/april/the-clean-hydrogen-future-has-already-begun.html.

[85]European Commission, Joint Research Centre, Green hydrogen opportunities in selected industrial processes, 26.06.2018, available at: http://publications.jrc.ec.europa.eu/repository/bitstream/JRC114766/kjna29637enn.pdf.

[86]National Grid's 2017 Future Energy Scenarios, available at: http://www2.nationalgrid.com/WorkArea/DownloadAsset.aspx?id=45609.

alone will not likely to be enough. There will be need for blue hydrogen too in order to kickstart the hydrogen market. Some Asian countries, including Japan, Korea and China, but also EU countries like Germany, are already producing personal mobility vehicles with hydrogen fuel cells. All electrification is not ideal from a system perspective as there is a need for flexibility with more links and storage.

Although it is still early days for hydrogen, there are new and promising projects for both small- and large-scale hydrogen projects including CCUS. Examples include from Frankfurt and Freiburg where hydrogen is integrated in the natural gas grid. In the UK, power to gas produced very close to the consumer is also taking place. There is also biomethane in the UK. Indeed, in the UK, there has been large-scale steam reforming taking place but other countries are focusing on electrolysis. Hydrogen can also be used for seasonal storage and the Dutch are looking into Blue Hydrogen with CCS and Green Hydrogen with renewable energy. Netherlands is not yet at the large scale utilisation stage, but industry is calling for a European innovation agenda to address the point that the cost is too high just now. Emphasising the growing awareness, the Netherlands now has a dedicated hydrogen envoy.

Equinor, Gasunie and Vattenfall are working on Blue Hydrogen with natural gas (SMR with CCS), the Magnum Project in the Netherlands and for this large-scale project, they are expecting to make an investment decision by 2023 and start production by 2028. Some other hydrogen projects and studies include Cadent Gas and Leeds H21 projects in the UK, and standardisation by CEN and Marcogaz conclusion on H2 injection into the gas transportation network. Europe is leading the industrial production of anaerobic digestors and electrolysers, with the development of renewable gas and decarbonised gas being perceived as good for growth and jobs.

Currently hydrogen is used mainly in industrial processes, some 55% of the hydrogen produced around the world is used for ammonia synthesis, 25% is used in refineries and around 10% is used for methanol production, largely based on (steam methane reformation—SMR), known as Grey Hydrogen. Around 68% of all of the hydrogen used in Europe is produced by steam reforming of methane. One can produce renewable energy only for hydrogen but again the economics are still not there (as fuel costs are high). Blue hydrogen could be SMR or CCS and now there are also Thermal Methane Pyrolysis (TMP) options, producing hydrogen and solid carbon.

Some of the most important challenges to hydrogen include: the scaling-up of hydrogen production; the blending of hydrogen into the natural gas network; and the competitiveness of biogas and hydrogen. The conversion of distribution networks to hydrogen is possible (although it has a cost). Hydrogen will be costly and a market has to be created. Important questions needed swift resolution include how it is going to be regulated? Who is going to do what? How to define the guarantees of origins? In relation to that, there are other questions such as whether gas sector decarbonisation should be a government responsibility or industry?

3.1.6.2 Biogas

Another green alternative to natural gas is biogas, which refers to gas produced by making use of biomass, such as manure, waste and other products or landfill and silt gas. This type of gas is also referred to as renewable gas. Biogas has historically been used in electricity production, heating and cooking and recently as a replacement to natural gas. In order to be injected into the natural gas system biogas has to be purified and upgraded to natural gas quality—referred to as biomethane. Biomethane could contribute to security of supply in the EU and play a role in energy transitions. However, new technologies are needed to facilitate the safe injection of biomethane into the existing natural gas system. In the EU, legal innovations may be necessary to facilitate this including on cross-border transport and trade of this energy source.

The Clean Energy Package does not specifically discuss biomethane but it is relevant as part of the revised Renewable Energy Directive.[87] The production of biogas is governed under Animal By-Products

---

[87]Directive (EU) 2018/2001 of the European Parliament and of the Council of 11 December 2018 on the promotion of the use of energy from renewable sources, OJ L328/82, 21.12.2018.

Regulation[88] if produced by landfill and fermentation, and waste regulations also have to be taken into account. The Renewable Energy Directive applies to biomass; however it does not provide a definition of biogas (although the proposed amendments introduced a definition albeit not detailed). In the absence of clearly defined regulations on guarantees of origin, gas quality, safety and sustainability criteria, currently different Member States, such as Denmark, Germany and the Netherlands, adopted different solutions. For effective integration of these green alternatives into the gas grid, more coordinated action needs to be taken at the EU level.

All in all, hydrogen from electricity is at an early stage. Synthetic natural gas from gasification also remains in its infancy but many biogas and small-scale biomethane plants are already operating. That being said, biogas could not replace current gas demand. Some form of CCUS is indispensable for the sustainable future of natural gas. It is preferable to produce hydrogen from gas and capture the CO_2, and it is also possible to use gas and take out the carbon downstream with retrofit gas plants. Further, methane leakage risks are mainly due to badly managed projects: with technology low cost and project management it can be avoided. It is important to note that hydrogen can be used for several synthetic fuels such as methane, dimethyl ether and diesel jet fuel. As renewable gas has small potential, it should be used in transport first. The European Association of Gas Turbines also noted that gas turbines can be adapted to new gas (biogas, biomethane, hydrogen) and that Combined Heat and Power (CHP) could produce renewable heat and electricity efficiently.

3.1.6.3 Carbon Capture, Utilisation and Storage

The energy transition is also a transition in manufacturing and it is happening now and CCUS is an example of this. It is particularly important for the decarbonisation of the industry. However, some consider

---

[88]Regulation (EC) No. 1069/2009 of the European Parliament and of the Council of 21 October 2009 laying down health rules as regards animal by-products and derived products not intended for human consumption and repealing Regulation (EC) No. 1774/2002 (Animal by-products Regulation).

decarbonisation as not the really the best way to describe CCUS. It is rather best described as a climate mitigation measure. There is a difference between CCS and CCU. For CCU there needs to be a market for the downstream utilisation of the carbon. The 2005 Intergovernmental Panel on Climate Change Special Report on CCS defines it as:

> Carbon dioxide (CO_2) capture and storage (CCS) is a process consisting of the separation of CO_2 from industrial and energy-related sources, transport to a storage location and long-term isolation from the atmosphere. This report considers CCS as an option in the portfolio of mitigation actions for stabilization of atmospheric greenhouse gas concentrations.[89]

In the US CCS is known as sequestration because it is not storage per se, as the carbon is not stored to be taken back later (or it could be defined as disposal but it is not a waste). The capture of CO_2 could take place pre- or post-combustion. If it occurs after electricity is produced for instance this would be post-combustion. Pre-combustion capture takes place in the process of gas treatment, and it can also be as oxyfuel via the steel industry.

In pre-combustion, methane is transformed into hydrogen and CO_2 is separated while the hydrogen is burned. Pre-combustion capture is the same technology that is used to make ammonia. In oxyfuel, instead of burning with air, the methane is burned with oxygen—and pure CO_2 can be added up post-emission capture. Oxyfuel combustion is one of the leading technologies considered for capturing CO_2 from power plants with CCS. Both pre- and post-combustion measures will be necessary going forward. CCS can be also part of Enhanced Oil Recovery (EOR). Additionally, there is bio-CCS, the concept of Bioenergy with CCS. Biomass with CCS would also result in negative emissions. Methane Cracking is another method to store CO_2 as black carbon. Both CCS and CCU could be short, medium and long term (in the long and medium term it could be used in building materials and as biofuels in the short term). The abatement success of CCS is highest in gas and chemicals, which is why the main consumer of hydrogen is

---

[89]Report IPCC, 2005—Bert Metz et al. (eds.) (Cambridge University Press, Cambridge, 2015).

currently the petrochemicals industry. Direct air capture of CO_2 uses is still not clear, but it cannot be done in cities.

CO_2 can be stored in an empty gas field, which is not really empty as some methane would still remain there. It works with the use of very thick layer of rock salt on sandstone and methane can be been trapped for millions of years. Scientist cannot guarantee complete safety but they provide a high probability of 99% + that the CO_2 would remain for hundreds of thousands of years. Aquifers have more capacity but the risks are uncertain and anyway there are not many suitable such sites in the EU—and the EU is more familiar with gas fields. Norway and the UK has the highest capacity for storage, with current CCUS capacity available in the EU to cover approximately 60 years of CO_2 emissions.

Globally it is China leading the technology, although in Australia there is a big project with Chevron to store CO_2 in deep aquifers. CCS however requires energy and that has a cost. For instance, in the case of power plants with CCS, there is a need of one fifth of the power produced for the CCS as significant amount of steam is required, and steam has a value. The CCS plant is also very material intensive, particularly steel, because of the required height and massive scale of the infrastructure, and to cope with the heat and steam.

Despite significant progress in decarbonising the power sector, GDP growth is translating into using resources more intensely. The IEA confirms that CCUS is crucial for big industrial emitters which contribute 24% of CO_2 emissions. Experts note that excluding CCS from the mix could increase climate mitigation costs by 138%. While there is a natural reduction of CO_2 contribution by the industry as a result of technological advancement, there simply has to be more innovation.

Investment in large scale CCS is actually still very modest, however there are some promising developments as the cost of CCUS projects are steadily becoming economic. CCUS technology is also not new. In the Netherlands early CCUS opportunities were identified in the 1990s. In 2005, CO_2 was sent through the pipelines in Rotterdam. Between 2012–2017 there was a Road project in the North Sea for offshore storage of CO_2 from coal plants although it was subsequently stopped. The Netherlands is one of the pioneer countries in exploring this technology in the EU, with the Dutch Government confirming CCUS subsidy for the industry in 2019.[90]

The transport and storage of CO_2 via pipelines has been done for decades especially in the US. Shipping CO_2 via sea routes is possible and can be done for long distance transport, as the majority of the costs lie in the capture and not on transport. In the UK, the CCUS Cost Challenge Task Force made a series of recommendations in July 2018 on how government and industry should work together to deploy CCUS in the UK by the 2030s.[91] Every scenario developed by the UK Committee on Climate Change to meet the carbon reduction commitments projects the continued use of natural gas and thus CCUS technologies could have an important role in reducing the carbon impact of natural gas, extending the life of existing plants and enabling clean industrial growth.

In the EU, the EU Taxonomy on Sustainable Finance referred to CCS as a key technology and included in all the pathways of the EU. The report states that *"CCS can be eligible in any sector/activity if it enables that primary activity to operate in compliance with the threshold—for example, steel, cement or electricity production".*[92] There are quite a number of ongoing projects in the EU. In 2017 and 2018 many CO_2 projects were considered as a PCI such as the Northern Lights, Rotterdam, Sapling, Teesside, ERVIA and Athos and TransPorts.[93] There are also other technologies such as the use of synthetic fuels in heavy duty vehicles, shipping and aviation. In a plant newly built in the Netherlands, steam is used in the winter for heating and used in the summer for CCS. Current plants are perhaps small but they present a learning

---

[90]The Dutch Carbon Capture, Utility and Storage subsidy schemes can be accessed via: https://www.rvo.nl/onderwerpen/duurzaam-ondernemen/energie-en-milieu-innovaties/carbon-capture-utilisation-and-storage.

[91]Carbon Capture Utilisation and Storage Cost Challenge Task Force, Delivering Clean Growth: CCUS Cost Challenge Taskforce Report, July 2018, available at: https://assets.publishing.service.gov.uk/government/uploads/system/uploads/attachment_data/file/727040/CCUS_Cost_Challenge_Taskforce_Report.pdf.

[92]Financing a Sustainable European Economy, The EU Taxonomy Technical Report, 18.06.2019, available via: https://ec.europa.eu/info/publications/sustainable-finance-teg-taxonomy_en.

[93]The full list is available here: https://ec.europa.eu/info/sites/info/files/detailed_information_regarding_the_candidate_projects_in_CO2_network_0.pdf.

curve, which is perfect for stepping up to bigger units. This plant costed around €20 million, whereas a large scale project could cost at least €100 million. The costliest part of this technology is the reforming section. Overall, the increased uptake of these technologies is largely because CCS is increasingly considered as a viable, flexible, complementary and indispensable technology.

Larger scale projects in the EU include H-vision in Rotterdam, a Blue Hydrogen project with participant including leading European energy companies such as BP, EBN, Engie, Equinor, Gasunie, GasTerra and Shell among others.[94] In Port of Rotterdam, Gasunie and EBN are jointly preparing a new project, Porthos, in which CO_2 generated by industry in Rotterdam's port area is captured and stored in empty gas fields deep under the North Sea seabed. The first CO_2 injection is planned for 2023.[95] A similar project is underway in the Port of Amsterdam, called Athos, which is close to big gas fields. They have 7 years to build the infrastructure, which may seem long but CCS is not a short-term solution. Between the Athos and Porthos projects depleted pipelines could be used, and the CO_2 could be used as a feedback which would establish a circular situation. These initiatives are a result of political back up, structural subsidy funds and an investment culture.

There are also a number of projects that use carbon (utilisation). Tata Steel uses oxyfuel, the Hisarna steel power project in the Netherlands captures CO_2 from coal also with the use of oxygen. It can be used for the production of sustainable building materials (CO_2 could be used instead of heat as a building element). Granulate can also be used to capture the CO_2. Furthermore, CO_2 can be injected into cement for use in the drying processes. Furthermore, in Iceland CO_2 is recycled via vulcanol to drive cars. Vulcanol is the brand name for renewable methanol produced from carbon dioxide and hydrogen from renewable sources of electricity (hydro, geothermal, wind and solar).[96] Synthetic kerosene for aviation fuel is being tested by KLM,

---

[94]More information about the H-vision project is available via: https://www.deltalinqs.nl/h-vision.

[95]More information about the Porthos project is available via: https://www.rotterdamccus.nl/en/.

[96]More information on vulcanol is available here: https://www.carbonrecycling.is/vulcanol.

and the Rotterdam airport area will also use direct air capture and solar to produce synthetic kerosene.

It is an incredibly fascinating time for expanding the limits of technology. All these initial initiatives will require funding to further develop and at the EU level the new Innovation Fund is foreseeing CCU funding.[97] A main barrier to some of these projects is that they use green energy to produce CO_2 as a feedstock and there is not a lot of it in the EU. Therefore, the scalability of zero carbon alternatives remains questionable. By looking at the existing proven technology, Blue Hydrogen is essential to replace natural gas, in particular in hard to decarbonise sectors of the economy. While there do exist tangible investments there will be the need for public support to bring prices down and overcome challenges and barriers. The regulatory framework in the EU does not yet reward CCS projects but this may change depending on the final version of the 2020 gas package.

3.1.7 Decommissioning

It would not be possible to conclude this chapter without discussing ongoing and planned decommissioning of natural gas infrastructure in the EU as this is linked with the future uses of such infrastructure in the transport of CO_2, hydrogen, syngas and renewable gases. In the case of the North Sea assets, studies point out that a significant proportion of existing assets in the energy sector have reached or are reaching their late asset lifecycle stage.

The burden and cost of decommissioning is expected to reach proportions that were (surprisingly) not foreseen. Therefore, the UK Maximising Economic Recovery (MER) Strategy recommends opportunities for the continued use of the infrastructure to be considered before the actual decommissioning of infrastructure.[98] Reusing and

---

[97]The Innovation Fund is one of the world's largest funding programmes for demonstration of innovative low-carbon technologies. See more at: https://ec.europa.eu/clima/policies/innovation-fund_en.

[98]Gokce Mete et al. The Maximising Economic Recovery Strategy for the UK, in Geoffrey Wood and Keith Baker (eds.), *The Palgrave Handbook of Managing Fossil Fuels and Energy Transitions* (Palgrave Macmillan, 2019).

repurposing also represents a significant opportunity to facilitate the success of the energy transition. If implemented gradually, transition policies would allow companies to adopt their business models and existing infrastructure. It will however, in some cases, not be possible to re-use or re-purpose the available hardware. In these cases the wells will be abandoned and the hardware will be dismantled. Where possible though the hardware should be recycled.

There are a number of instruments of international law that potentially apply to decommissioning of offshore pipelines, including the Convention on the Continental Shelf and the Law of the Sea (UNCLOS). The London Protocol on 'dumping', the Convention for the Protection of the Marine Environment of the North-East Atlantic (OSPAR) Decision (1983) and the IMO standards also come into play. The problem with the existing frameworks is that their provisions do not reflect the current situation. For instance, the Continental Shelf Convention requires installations to be removed entirely after their end of life. However, today, we know that it is better to leave some installations, either partially or entirely. A further pause for thought is that abandonment of wells has a high cost, and re-use or re-purposing mitigates this.

These instruments mention installations but it should be pointed out that this does not automatically cover pipelines and wells as they indicate platforms. Today there are over 40,000 kilometres of pipelines in the North Sea and it could cost around £60 billion to decommission them, which will eventually be reflected in taxes. There will be naturally occurring radiation, sometimes they are covered by rocks or a concrete mattress to stabilise the overall structures, and they may also often be covered by marine growth, raising the question of whether decommissioning protects or damages the environment by removing the pipelines.

Although there is limited state experience in the world on decommissioning, the UK has the most advanced regime but somewhat paradoxically there is no regulation only Oil and Gas Authority guidelines. Decommissioning in the UK is now led via guidance documents issued by the Oil and Gas Authority, the UK economic regulator of the sector. As per the guidelines, the pipeline operator prepares a plan and carries out a comparative assessment of different options—for example, whether it should be removed or not. The problem with decommissioning in the UK is that there is a residual liability which causes concern amongst industry. The company that placed the pipelines in place may not be around anymore, the assets might have been divested, although the initial company can always be called back as per this residual liability. In Norway, the state can take over this residual liability against a one-off payment. In the UK, the tax regime did include decommissioning allowances but companies could offset them against their profits. If at the end of the decommissioning the treasury saved money, it would have to give it back to the industry.

The final issue is public acceptance. There are millions of people employed in the fossil fuels sector, therefore it is important to focus on low hanging fruits and minimise the impact of these costs being placed on the tax payer. The use of these pipelines and wells for CCUS instead of decommissioning could prove to be an ideal solution. The prospect of the reuse of pipelines also offers some solutions to deal with stranded assets. However, when investing today in new pipelines, because it might be used by other gases in the future, each TSO will need to make a relevant assessment for hydrogen because steel quality varies.

In terms of policy and regulation in the EU, the 2019 New Guidance on how to consider the environment in oil and gas extraction[99] includes Best Available Techniques for decommissioning which it describes as the removal of facilities and remediation of a site used for the production of hydrocarbons. It usually refers to offshore facilities, and indeed the Guidance Document only considers decommissioning for offshore facilities. This guidance document also recommends consideration of the reuse of infrastructure prior to decommissioning and may include use by the hydrocarbons industry as well as other sectors. Perhaps a missed opportunity, the EU gas package 2020 is not expected to provide further provisions on the decommissioning of natural gas infrastructure in the EU.

---

[99]European Commission, Best Available Techniques Guidance Document on Upstream Hydrocarbon Exploration and Production, 27.02.2019, available at: https://ec.europa.eu/environment/integration/energy/pdf/hydrocarbons_guidance_doc.pdf.

3.1.8 2020 Gas Package

In the EU energy regulatory space, any changes to electricity market legislation are often followed by changes to the gas market. The text of the revised Electricity Directive is adopted in March 2019.[100] While there is no official announcement as of July 2019 about the possible recast gas directive, many DG Energy officials, other EU bodies on energy (such as ACER) and industry associations have already started engaging in a debate on what to expect from the 2020 gas package (this may not be what it is finally called of course, but it appears to be the name given by the Brussels circles for the time being). That said, the DG Energy Deputy Director General noted in an interview that is could be named the Decarbonisation Package. Eurogas, an industry association of gas wholesale, retail and distribution companies would like it to be called the Sector Integration Package, to point out market coupling between electricity and gas.

Not much is publicly announced as there will also be a new Commissioner for energy in late 2019. With this package, the new Commission will decide on new policy orientations which again are not fully known as of yet.[101] At the final stages of this book, the new Commission President Ursula von der Leyen released the European Green Deal as part of her agenda for 2019–2024. The European Green Deal will have an impact on the upcoming gas package. Ostensibly, one can expect extensive consultations, policy proposals and negotiations on the gas market as occurred for the electricity sector four years ago. In fact, ACER has already launched a public consultation to gather stakeholders' views and further information regarding trends in the European energy sector—and particularly in the gas sector—beyond 2025. The input from the consultation will be used by the Agency to prepare recommendations to the European Institutions on possible

---

[100]European Parliament legislative resolution of 26 March 2019 on the proposal for a directive of the European Parliament and of the Council on common rules for the internal market in electricity (recast) {COM (2016) 0864—C8-0495/2016—2016/0380 (COD)}.

[101]Political Guidelines for the Next European Commission 2019–2024, available at: https://ec.europa.eu/commission/sites/beta-political/files/political-guidelines-next-commission_en.pdf.

future legislation.[102] Realistically, it is likely that the new gas legislation would be published either in late 2020 or even in 2021.

It can be presumed that in the shorter term the legislation will support natural gas, however in the long term fossil fuel gas will have to be replaced with renewable gases and the gas market has to be reformed and methane emissions have to be drastically reduced on a lifecycle basis by 2050. Short-term, gas market reforms may focus on tailor made regulation that would show a clear split within Europe, between North West and South East Europe, and remedy markets in some regions via capacity and commodity release programs and market mergers. Licensing schemes for supply and trading are currently national and widely different, so these can be reformed to harmonise across borders. LNG is also treated differently across the EU Member States so new provisions may target EU wide synchronisation of LNG regulation. There is also an issue with tariff pancaking (mentioned in Chapter 1), that the entry-exit tariffs means that market players pay twice, when entering and exiting the network. Inter-TSO mechanisms may be necessary to resolve this issue.

In the long term, it is becoming increasingly difficult to see natural gas without any form of CCS/CCU, therefore the regulation could enable alternatives to natural gas to be identified in a timely fashion. Sector coupling between gas and electricity would also likely be covered. Sector coupling originally referred to the electrification of end-use sectors like heating and transport, with the aim of increasing the share of renewable energy in these sectors and as a balancing tool. However today it is also used to capture supply-side sector coupling which focuses on further integration of gas and electricity sector through technologies such as power to gas (i.e. Green Hydrogen). Indeed, there will be provisions covering hydrogen but it is not clear which technology will be included within its scope, Blue, Green or Grey Hydrogen? Ideally, first there needs to be an impact study for instance on how much hydrogen can be transported via existing pipelines and whether there is a need for new infrastructure for instance for biomethane, oxyfuel or hydrogen, or

---

[102]ACER Public Consultation the Bridge beyond 2025 is, available at: https://www.acer.europa.eu/Official_documents/Public_consultations/PC_2019_G_06/The%20Bridge%20beyond%202025%20-%20PC_2019_G_06.pdf.

perhaps it could be decided that hydrogen will be a local option in the industry only. This is important as gas quality is still not harmonised in the EU. The natural gas industry associations have been demanding the introduction of an EU target for renewable and decarbonised gases (as high as 10%). There are calls for taxonomy of gas to demonstrate climate credentials, while the burden of proof should fall on the industry. They would further like to see the EU to develop a blueprint for guarantees of origin for hydrogen and a market design to enable renewable and decarbonised gases to reach maturity and development at scale and to be integrated in the wholesale market.

The new legislative proposals will surely go hand in hand with market upgrading and modernisation. New products of TSOs will emerge and implementation of network codes will be accelerated. Inevitably this exercise will aim at enabling the development of technologies to decarbonise gas supply, however, from the outset, the Commission does not sound very happy about the subsidies (despite the fact that they are continued to be widely used to fund natural gas [security of supply] infrastructure). The Commission's view is that competition should choose the right technologies towards decarbonisation, as they believe that policy makers are not the best ones to choose technologies. However, DG Climate has indicated support for renewable and low carbon gases in a similar form that is provided for renewables which were able to penetrate into the market thanks to the regulatory push.

On the decarbonisation of the gas sector, it should ideally identify the roles of different actors, the TSO, DSO and supplier. The package could also cover investment aspects and regulate pathways for socialising costs. This is also applicable for CCUS, the gas package could decide whether it should be regulated—therefore costs would be socialised or left to the market.

Finally, as security of supply challenges have changed since the Third Energy Package was drafted in 2009, so these will need to be renewed and new challenges may be added.

3.1.9 Chapter Conclusions

This Chapter concludes that gas infrastructure will continues to play a significant role in making European energy consumption more sustainable. However, coal-to-gas switching benefits will be limited beyond 2030 and unabated gas cannot continue beyond 2050. While in the mid and short term the continuous indigenous gas production decline will make further natural gas imports necessary, the climate-neutral energy system of the future appears to have no room for an entire sector that is traditionally oriented toward the fossil fuel natural gas.

Decarbonisation of gas, while not a new concept, may be more difficult than expected. However, it is necessary as important sectors of the economy like heating, mobility and the industry cannot be sustained with an all electricity system in the future. The only option for offering the same flexibility of natural gas to the system at scale currently appears to be blue hydrogen solutions with CCS. That said, regulatory, market and technology challenges are substantial. Important questions such as who will operate power to gas plants remains unclear, as it is not defined whether this is energy production or conversion. Green hydrogen produced with excess renewable electricity for instance could be defined as conversion and in that case TSOs may not be in charge. However, if synthetic gas is produced and reinjected to the grid this would be production, yet the TSO may still not be in charge because of EU unbundling requirements.

Blue hydrogen and CCS are considered inevitable for the industry, and large-scale decarbonisation of the transport sector via hydrogen (particularly large trucks). The timeline of these initiatives and their financing remains uncertain. Many questions have been raised as to who will pay for hydrogen; these too are also left unanswered today.

Future work on the 2020 gas package will need to adopt effective definitions on what can be regarded as green or low carbon but at the same time fundamental data should be gathered on the compatibility of the current infrastructure with these new sources. The next generation of gas legislation in the EU should allow the market to commit to investments where it is needed but also be technology neutral for efficient outcomes. Finally, proper incentive schemes should be put in place and lessons learned from natural gas market development, in particular with subsidies and public funding as the issue of where tax payers money will be spent is becoming more important than ever before. The next chapter will explore historical trends in financing natural gas infrastructure, covering existing public and private funding structures, in order to draw some lessons for the future of gas in the EU.

4

Mapping Natural Gas Subsidies and Natural Gas Project Finance in the Light of Energy Transitions in the EU

4.1 Introduction

Energy transitions refers to the gradual switch from a fossil-based energy system to a sustainable energy one, fuelled by clean energy sources. While renewables will play a crucial role in decarbonising the energy industry their contribution may not come in time to keep within the carbon budget left to be emitted into the atmosphere defined by the 2 °C target. Previous discussions were often based on an implicit assumption that only renewable energy is carbon free energy. However, as previous chapters of this book have already shown, hydrocarbons can be decarbonised with disposal of the resulting CO_2 in geological structures or using the resulting carbon as Black Carbon. One the one hand, natural gas is best suited for decarbonisation and could contribute to large-scale decarbonisation of the energy sector. On the other hand, decarbonising natural gas would maintain the value of gas resources and of investment in the gas industry.

This is important as currently all of the natural gas infrastructure projects selected as Projects of Common Interest in the EU

© The Author(s) 2020
G. Mete, *Energy Transitions and the Future of Gas in the EU*, Energy, Climate and the Environment, https://doi.org/10.1007/978-3-030-32614-2_4

are supported by public funds as they are not commercially viable. Subsidising fossil fuels undermines climate change, energy security and sustainable development objectives. Fossil fuel subsidies also put a restraint on public budgets, limiting the funds that can be spent on economic diversification, education and health alike, and they also distort competition with low carbon energy technologies. Therefore, fossil fuel subsidy reform is an important policy issue for all countries.

This chapter will provide an overview of public funds created to support the development of gas infrastructure, including LNG import facilities, storage and pipelines (both interconnections and third county pipelines) in the EU. It will use case studies to reveal the role of subsidies and the future of such subsidies in light of the Clean Energy Package objectives. As the prospects for decarbonising the gas network can play an important role as part of a realistic and long-term energy transition strategy in a gradually evolving low carbon economy, this chapter will also examine how gas infrastructure could be financed on a commercial basis without recourse to public subsidies and what the possible impact of the sustainable finance package will be on the gas industry's ability to source funding from investment entities.

Finally, it will analyse the impact of the carbon price and the improved performance of the EU ETS on company and government policies in their decisions to invest/support gas infrastructure (as opposed to renewable energy infrastructure). Carbon pricing could have a significant impact on the speed of energy transition and profitability of fossil fuel industries. Carbon pricing (along with removal of inefficient fossil fuels) makes low carbon solutions, such as renewables and carbon capture, use and storage, more cost competitive. This chapter will conclude with a discussion on lessons learned from the traditional methods of financing gas infrastructure and raise the question as to what innovative alternatives are available for the future of (decarbonised) gas infrastructure (e.g. financing of CCUS, biomass, hydrogen).

4.2   Public Funds

In the previous chapter, the general rule for development of gas infrastructure in the EU was explained. According to Articles 9–11 of the Gas Directive, natural gas infrastructure should be developed to allow non-discriminatory third-party access and the pipeline should be unbundled; the owner of the natural gas cannot be the owner and operator of the pipeline infrastructure. This general rule has been in place since the 2003s Energy Package. However, nearly all of the new cross-border pipeline projects to date have been developed based on the 'exemption' and not the rule. The exemption allows for derogations from implementation of mandatory third-party access for a defined and temporary time period for new investment in gas infrastructure projects within the EU. This is because the de facto rule clashes with the traditional principles of commercial project financing for long-term capital-intensive investments.

Understandably, liberalisation of energy infrastructure is a typical importing country policy. However, it has a cost. This is because opening up competition such as through mandatory third-party access inevitably prolongs the pay-back period of an investment and could therefore raise the capital costs of the project. This is particularly true in the case of pipeline projects which are high-risk and capital-intensive ventures. On top of this, European natural gas markets are moving from long-term natural gas markets to spot markets, while gas producers are invited to commit to long-term infrastructure investments. There are different time-frames characterising supply and demand. If the producers do not make the upstream investments, in the view of lack of finances or demand to construct the transport infrastructure, the gas market will experience a mismatch. This is part of the security of supply and ensuring environmental sustainability dilemma, and not an easy one to tackle. Unfortunately, shorter markets in the energy sector does not make an easy case for investments.

Therefore, the Commission has been topping up, with public funds, the financing for projects which can secure a certain amount of commercial funding. This increases the chances of the project being eligible for borrowing. For example, the TAP pipeline project, connecting Turkey, Greece, Albania, and Italy, jointly owned by BP (20%), SOCAR (20%), Statoil (20%), Fluxys (16%), Total (10%), E.ON (9%), and Axpo Trading (5%) has an estimated cost of US$1.8 billion. The feasibility studies for the project were partly financed by the EU public funds.[1] TAP and its shareholders Axpo Trading, Statoil, Fluxys and E.ON Ruhrgas signed a funding agreement with the Shah Deniz consortium for the pipeline project. In addition to this, EBRD, the EIB and the Export Credit Guarantee agencies of OECD countries, as well as commercial banks, provided funds for the project. The EBRD also provided US$500 million to fund the completion of the TANAP project that passes through Turkey.[2] The total fixed investment costs of TANAP is US$9.204 billion.[3] The remaining funds came from a syndication of commercial and financial institution loans (TANAP pipeline projects details are further debated in the next chapter). The next sections will provide a synopsis of EU funding sources that have been made available for interstate natural gas pipeline projects to demonstrate their magnitude and versatility.

4.2.1 The 2020 European Fund for Energy, Climate Change and Infrastructure (Marguerite Fund)

The first public financial source that will be covered in this chapter is the Marguerite Fund, a long-term pan-European equity fund. The Fund was established in 2010 to provide seed financing for key investments in capital intensive renewables, energy and transport sectors.

---

[1]EU Parliamentary questions, Answer given by Mr. Arias Cañete on behalf of the Commission, 31.08.2017, available at: http://www.europarl.europa.eu/sides/getAllAnswers.do?reference=E-2017-004756&language=EN.

[2]EBRD Press release, EBRD Board approves financing for Trans-Anatolian Natural Gas Pipeline, 18.09.2017, available at: http://www.ebrd.com/news/2017/ebrd-board-approves-financing-for-transanatolian-natural-gas-pipeline.html.

[3]News Article, Daily Sabah, The project benefits from incentives provided by Turkish Government, such as VAT exemption, exemption from customs tax and employer grants for insurance premiums, 27.02.2015, available at: http://www.dailysabah.com/energy/2015/02/27/tanap-will-be-supported-with-incentives.

In the energy sector, core areas of investment include electricity and gas transportation and interconnection infrastructure, such as pipelines and high voltage transmission lines. The fund presents a hybrid approach to financing by combining the market-based principle of return to investors with the objective of achieving public policy goals. The fund's total commitment is €710 million, of which €80 million comes from the Commission with the rest being distributed among other investors. These include six core sponsors that are major European financial institutions: EIB, Caisse des Dépôts et Consignations, Cassa Depositi e Prestiti, Instituto de Crédito Oficial, Kreditanstalt für Wiederaufbau, PKO Bank Polski. In 2016, the Marguerite Fund acquired a 28.97% stake in the Latvian natural gas company, AS Latvijas Gāze, which owns the third largest gas storage facility in Europe, a facility of significant importance to security of supply.[4]

4.2.2   TEN-E Trans-European Energy Networks (2009–2013)

The Trans-European Networks (TEN) fund was created to link network bound sectors, railway, energy and telecommunications in the EU[5] over the 2007–2020 period. In reality, The TEN scheme only ran between 2007 and 2013. The Commission calculated the cost of completing the TEN as €600 billion and €160 billion was allocated until end of 2013.[6] The Trans-European Networks for Energy (TEN-E) strategy was focused on connecting the energy infrastructure in Member States. Priority corridors with urgent need of infrastructure development, covering at least two Member States, were established. Priority gas corridors

---

[4]Press release, The Marguerite Fund Completes the Acquisition of a Stake in AS Latvijas Gāze in Latvia, 28.01.2016, available at: http://www.marguerite.com/2016/01/the-marguerite-fund-completes-the-acquisition-of-a-stake-in-as-latvijas-gaze-in-latvia/.

[5]Regulation No. 680/2007 of the European Parliament and of the Council of 20 June 2007 laying down general rules for the granting of Community financial aid in the field of the trans-European transport and energy networks.

[6]Communication from the Commission—Trans-European networks: Towards an integrated approach {SEC (2007) 374}/COM/2007/0135 final.

in the gas sector mainly focused on diversification and security of supply and include: North-south gas interconnections in Western Europe, North-South gas interconnections in Central, Eastern and South Eastern Europe, Southern Gas Corridor and the Baltic Energy Market Interconnection Plan. In addition, a number of domestic gas pipelines have been funded by the facility, especially in Poland and Greece.[7] In 2014, the TEN-E programme was replaced by the Connecting Europe Facility (CEF) to cover the period until 2020.

4.2.3 Connecting Europe Facility (CEF)

The CEF is an EU fund to support trans-European networks, including transport, energy and digital services. The aim of CEF is to act as a catalyst for further private and public-sector investment in key infrastructure projects that promote growth, jobs and competitiveness. Between 2014 and 2020, €5.35 billion has been dedicated to the energy sector.[8] In 2018, a further €200 million of CEF funding was announced to be made available for energy projects including gas infrastructure.[9] CEF funds are used to facilitate the inception of PCIs that are identified across Europe. The energy budget responds to energy security and system resilience concerns following the 2009 Russia–Ukraine crisis, as this created awareness of the need for more interconnection and diversification of capacity. Increased interconnectivity with reverse flows enables the EU to manage any supply crisis or shortfall

---

[7]Database for investment through EU regional policy programmes, available at: http://ec.europa.eu/regional_policy/en/projects/ALL?search=1&keywords=&countryCode=ALL&regionId=ALL&themeId=ALL&typeId=ALL&progperiod=2&dateFrom.

[8]More information on CEF facility is available at: https://ec.europa.eu/inea/en/connecting-europe-facility/cef-energy.

[9]EU Commission News, €200 Million EU Funding for Cross-Border Energy Infrastructure, 19.03.2018, available at: https://ec.europa.eu/info/news/eu200-million-eu-funding-cross-border-energy-infrastructure-2018-mar-19_en?pk_campaign=ENERNewsletterApril2018.

more effectively. The Commission expected that the CEF *"will reduce risks and increase effectiveness and efficiency of EU funding"*.[10]

In 2014, under the first call for proposals for CEF funding in energy, 34 projects were offered €647 million of financial support in total.[11] The then EU Commissioner for energy noted regarding the CEF that:

> *To be absolutely clear, I must emphasise that words are not enough to improve the EU's infrastructure. What we need is action, and this demands a greater effort on all sides. It means a stronger commitment by project promoters when examining technical ways accelerating implementation; by regulatory authorities who must agree on cost allocation and funding; and by national ministries that need to ensure the necessary political support while also tapping into all possible financing sources, including the EU's structural funds, cohesion funding and the European Investment Bank.*[12]

In 2019, CEF call proposals reached a total of €750 million and focused on, among others, increasing competitiveness by promoting the integration on the internal energy market and the interoperability of electricity and gas networks across borders, enhancing security of supply and eliminating bottlenecks in the energy system. It also covered the development cross-border carbon dioxide networks. The EU has been funding CCS demonstration plants via the CEF, and the Port of Amsterdam in the Netherlands and Northern Lights Projects in Norway, as well as a hydrogen transport hub in the north of England are among those projects.

In principle, the CEF applies only to projects that have clear European benefits and that cannot be developed on purely commercial

---

[10]Directorate-General for Internal Policies Policy Department: Budgetary affairs assessment of Connecting Europe Facility in-depth analysis, 19.01.2016, available at: http://www.europarl.europa.eu/RegData/etudes/IDAN/2016/572677/IPOL_IDA(2016)572677_EN.pdf.

[11]European Commission, Press release, Energy: EU Invests €150 Million in Energy Infrastructure Brussels, 14.06.2015, available at: http://europa.eu/rapid/press-release_IP-15-5362_en.htm.

[12]Press release, Friends of Europe, Günther Oettinger: Energy Security Is Within Our Grasp if the EU Plays as a Team, 09.10.2014, available at: http://www.friendsofeurope.org/greener-europe/energy-security-is-within-our-grasp-if-the-eu-plays-as-a-team.

terms. For instance, the Commission invested €187.5 million for the Baltic-connector project, which is a planned strategic gas pipeline between Estonia and Finland to allow Finland to diversify away from its sole natural gas supplier, Russia.[13] The project is not a commercial one but implements the solidarity and security of supply principles of the EU. The funding comes from the CEF and covers around 75% of the total funding needed for the infrastructure project. The same approach was taken for supporting an earlier interconnector project between Poland and Lithuania. In this case, the Commission provided around €305 million under the CEF.[14] In December 2017, the EU Innovation and Networks Executive Agency, Polish operator Gaz-System and its Slovak counterpart Eustream, signed an agreement to construct a new gas pipeline connecting Poland and Slovakia. This project also received CEF funding worth €107.7 million.[15]

The CEF facility also provides funds for projects that involve non-EU countries, like Turkey. What is relevant for this study is that the CEF facility provided 30% of the total funding of the environmental monitoring costs of the TANAP project, worth €6.7 million in total, and funding for 50% of engineering studies for TANAP SCADA system, worth €5.1 million in total. There has also been €2 million funding from the CEF facility for the pre-front end engineering design studies of a potential East-Med pipeline. The CEF-E Gas platform contains the full list of PIC that are financed by the CEF.[16]

---

[13]EU Commission News, EU Invests 187.5 Million Euro in First Gas Pipeline Between Estonia and Finland, 10.08.2016, available at: https://ec.europa.eu/energy/en/news/energy-union-eu-invests-1875-million-euro-first-gas-pipeline-between-estonia-and-finland.

[14]European Commission, Fact Sheet, End of energy isolation in the Baltics: How the Gas Interconnector Poland-Lithuania (GIPL) works Brussels, 15.10. 2015, available at: http://europa.eu/rapid/press-release_MEMO-15-5845_en.htm.

[15]News Article, Central Europe Energy Partners, EU Grant Agreement for the Poland-Slovakia Gas Interconnection, 10.01.2018, available at: https://www.ceep.be/eu-grant-agreement-poland-slovakia-gas-interconnection/.

[16]CEF-E Gas platform is available at: https://ec.europa.eu/inea/en/connecting-europe-facility/cef-energy/projects-by-sector/gas.

4.2.4 Projects of Common Interest (PCI)

The PCI is a framework established by the Commission in 2013 to support strategic projects to create an integrated EU energy market. The framework specifically targets cross-border projects that can link national energy systems. These projects are identified every two years and the list is published by the Commission. The selection criteria are established under the Guidelines for TEN-E,[17] the European Structural and Investment Funds (see below) and the CEF, which replaced funding available under the TEN-E programme during the period 2007–2013. PCI benefit from accelerated licensing procedures, improved regulatory conditions and access to financial support.

In order to qualify as a PCI, a project must:

- involve and have a significant impact on at least two EU Member States;
- enhance market integration and contribute to the integration of Member States' networks;
- increase competition in the energy markets by offering alternatives to consumers;
- enhance security of supply; and
- contribute to EU energy and climate goals, which should facilitate the integration of an increasing share of energy from variable renewable energy sources.

On the 14th October 2013, the European Commission adopted a list of 248 key PCI infrastructure projects, in electricity and gas transmission, storage and LNG, as well as in smart grids and in oil.[18] There were 107 natural gas projects in this list, predominantly from Central and

---

[17]Regulation (EU) No. 347/2013 of the European Parliament and of the Council of 17 April 2013 on guidelines for trans-European energy infrastructure and repealing Decision No. 1364/2006/EC and amending Regulations (EC) No. 713/2009, (EC) No. 714/2009 and (EC) No. 715/2009.

[18]Information on CEF Energy Actions are available at: https://ec.europa.eu/inea/en/connecting-europe-facility/cef-energy/cef-energy-projects-and-actions.

South East Europe, and of those gas projects 12 related to the Southern Gas Corridor aimed at increasing gas imports to the EU.[19] A new list is produced every three years from which completed, unfeasible and abandoned projects are removed. The second and third lists were made public in 2015[20] and 2017, respectively; the second list had 195 key energy infrastructure projects and the latest included 173 projects.[21]

PCI projects can benefit from available funds such as the CEF, the European Structural and Investment Funds, Project Bonds and EBRD and EIB funds. However, not all projects listed will need support from public funds.[22] Projects could be built on a commercial basis. However, there are no cross-border natural gas pipeline projects listed as PCI which have not required financial assistance. Overall, around €70 billion is expected to be invested in gas projects. In principle, the EU's financial assistance should not exceed 50% of the total costs for feasibility studies and works—which can increase up to 75% in exceptional cases for projects that contribute significantly to security of supply, enhancing energy solidarity between Member States or offering highly innovative solutions.

The Southern Corridor, a mega-Euro-Caspian project is listed as a PCI, including the TAP, TANAP and TCP. The TCP aims to bring natural gas from Turkmenistan, which currently has no major natural gas export routes to the EU. Furthermore, the 180-kilometre-long Greece-Bulgaria interconnector, which could supply all countries that were supposed to be part of the South Stream with natural gas that flows from the Southern Corridor, is also included in the list.[23]

---

[19]The 2013 PCI selection is available at: https://ec.europa.eu/inea/sites/inea/files/download/publications/pci_ener_superfinal.pdf.

[20]The 2015 list of PCIs are available at: http://eur-lex.europa.eu/legal-content/EN/TXT/PDF/?uri=OJ:JOL_2016_019_R_0001&from=EN.

[21]The 2017 list of PCIs are available at: https://ec.europa.eu/energy/sites/ener/files/documents/annex_to_pci_list_final_2017_en.pdf

[22]One example is the AQUIND Interconnector which will link the British and French electric power grids. More information is available at: http://aquind.co.uk/.

[23]More information on the IGI Poseidon project is available at: http://www.igi-poseidon.com/en/igb.

There are also Projects of Energy Community Interest (PECIs) and PMIs, which involve non-EU but EC Member States.[24] Projects selected by the Energy Community Ministerial Council benefit from streamlined permitting procedures within Contracting Parties in cases where the Competent Authorities are put in place, and where applicable, from cross-border cost allocation.

4.2.5 European Structural and Investment Funds

The European Structural and Investment Funds (ESIF)[25] are managed by the Commission and the Member States and their main purpose is to increase employment and the sustainability of the economy and environment in the EU. The funds, which consist of the ERDF, European Social Fund, Cohesion Fund, European Agricultural Fund for Rural Development and European Maritime and Fisheries Fund support over half of the EU funding budget. Their investment areas cover, inter alia, the Energy Union and Internal Market. ESIF funds committed around €69 billion for the period 2014–2020 to support the Energy Union targets.[26]

In order to avoid duplication with the CEF, Member States and the Commission ensure that ERDF and Cohesion Fund interventions are planned in close co-operation with the support provided from the CEF. ESIF priority areas for investment in the energy sector are concerned with projects that have an impact beyond a certain Member State, such as North-South gas interconnections in Central Eastern and South Eastern Europe. These projects include the Lwówek-Odolanów pipeline, which will connect the LNG terminal in the city of Świnoujście

---

[24]Selection process for priority projects of Energy Community is available at: https://www.energy-community.org/regionalinitiatives/infrastructure/selection.html.

[25]More information on The European Structural and Investment Funds is available via their homepage at https://ec.europa.eu/info/funding-tenders-0/european-structural-and-investment-funds_en.

[26]EU Commission News, Energy Union: The EU Cohesion Policy Invests to Improve Energy Security in Poland, 15.06.2017 http://ec.europa.eu/regional_policy/en/newsroom/news/2017/06/15-06-2017-energy-union-the-eu-cohesion-policy-invests-to-improve-energy-security-in-poland.

in North-West Poland to existing gas transmission, storage and distribution infrastructure. The project is identified as a PCI and could potentially supply Ukraine, the Czech Republic and Slovakia with natural gas. Furthermore, €112.5 million from the ERDF will be allocated for the construction of a 167 kilometre-long domestic natural gas pipeline in Poland.[27]

On ESIF, the European Commissioner for Climate Action and Energy, Miguel Arias Cañete, noted in 2015 that

> *the completion of a truly competitive EU-wide energy market is essential in order to turn the Energy Union into a reality. But without reliable and well-connected energy networks this will not happen. This is why we are investing in projects to integrate the market further and to diversify sources and routes, in particular in Central Eastern and South Eastern Europe as well as in the Baltic region.[28]*

Funding will also be focused on studies that aim to remove obstacles which obstruct the free flow of natural gas in the EU gas transmissions system.

4.2.6 Ten-Year Network Development Plan (TYNDP)

The selection of PCIs involves TSOs to submit an annual TYNDP to their NRAs, containing efficient measures guaranteeing the adequacy of the gas system and security of supply. These national TYNDPs are then synced to the EU-wide TYNDP, which are developed by ENTSOG.[29] The TYNDP helps to prioritise the limited projects which will receive public funds in support towards their realisation. The plan,

---

[27]More information on the ERDF is available at: https://cohesiondata.ec.europa.eu/funds/erdf.

[28]European Commission—Press release, Energy: EU Invests €150 Million in Energy Infrastructure, Brussels, 14.07.2015, available at: http://europa.eu/rapid/press-release_IP-15-5362_en.htm.

[29]Regulation (EU) No. 1316/2013 of the European Parliament and of the Council of 11 December 2013 Establishing the Connecting Europe Facility, amending Regulation (EU) No. 913/2010 and repealing Regulations (EC) No. 680/2007 and (EC) No. 67/2010, Official Journal of the European Union, L348/129.

which looks 20 years ahead, also aims to provide the right signals for future investment in pan-European gas infrastructure. This long-term assessment requires the development of demand scenarios. The 2017 TYNDP, taking into account the EU's ambitious climate and energy targets for 2030, reports that:

> *European indigenous natural gas production is set to decline in a number of countries, in particular in the Netherlands where the depletion of the Groningen field is under close monitoring by the authorities. In a context where achieving the EU climate targets could result from either an increase or decrease of gas demand by 2030, this implies that European supply needs are foreseen to increase or at best stay stable.*[30]

The report further notes that, while there had been progress since the 2015 TYNDP with the implementation of 19 projects directed towards completion of the internal market, there are challenges in attracting new suppliers for diversification due to uncertainty regarding the future of gas. As acknowledged by ENTSOG, the current gas infrastructure network is a result of many decades of planning and investment. Hence, forward-looking investment in diversified supply sources is considered key to ensuring security of supply in the event of a supply- or infrastructure-related crisis.

ENTSOG is responsible for developing a Cost-Benefit-Analysis (CBA) methodology for energy system wide analysis to support the PCI selection process which, as explained above, aims to facilitate investment in energy infrastructure in order to achieve the EU's energy and climate policy objectives. The CBA is designed to calculate the potential investment costs of projects and the benefits they bring to the energy market and the EU. The benefits, known as indicators, considered are; security of supply, social and economic welfare, integration of renewable energy sources, CO_2 reduction, impact on losses and technical resilience.

---

[30]ENTSOG, Ten-Year Network Development Plan 2017, available at: https://www.entsog.eu/public/uploads/files/publications/TYNDP/2017/entsog_tyndp_2017_main_170428_web_xs.pdf.

The uncertainty over gas in the EU energy mix results in a lack of market commitment by shippers. This adds to the current situation that none of the cross-border PCIs have been developed on a commercial basis without the support of public funds. The investment is only triggered by security of supply reasons. The TYNDP allows co-financing, but this carries the risk of further reduction of the network users' willingness to commit to the new infrastructure, as the EU funding is often seen as a source of delay or uncertainty.[31] On top of this, experts note that the EU co-sponsored projects in fact have a limited impact on competitive internal market objectives.[32]

Infrastructure utilisation is not the same for each infrastructure. It may be a result of some of the inefficiencies of the PCIs and CBA methodology which is not advanced enough. Most CEF funding currently goes to Central and South-Eastern Europe. In the future, there is a need to develop refined monitoring tools and improved CBA methodology, such as taking the lifecycle value of costs and the monetised benefits into account. National development plans typically include the national sections of cross-border gas transmission projects, but do not consider the cross-border effects of LNG or underground gas storage projects located outside their geographical scope. Year by year the number of PCIs are decreasing. It would be advisable to take a closer look at the current primary energy mix and the scale of changes needed to reach a carbon-neutral EU by 2050 which will be extremely challenging in all sectors.

4.3   Raising Funds on a Commercial Basis

There are two main ways of financing energy infrastructure projects in the EU. Electricity and gas investments could be developed on a commercial basis without subsidies, or with public finance, which could be

---

[31]Ten year network development Plan 2015, Entso-G, available at: http://www.entsog.eu/publications/tyndp.

[32]Special Report No. 16/2015: *Improving the Security of Energy Supply by Developing the Internal Energy Market: More Efforts Needed*, available at: http://www.eca.europa.eu/Lists/ECADocuments/SR15_16/SR_ENERGY_SECURITY-EN.pdf

in the form of seed money or borrowing, as covered above. It has also been explained that there are no natural gas pipeline projects in the EU that have been selected as a PCI and proceeded without subsidies. This does not mean that cross-border energy infrastructure projects are not suitable to be constructed on a commercial basis. There are examples of projects outside the regulatory sphere of the Third Energy Package that have been developed on a commercial basis. Commercial energy projects could be developed based on corporate financing, shareholder financing or project (debt) financing, or a mix of them.

4.3.1 Corporate Financing and Shareholder (Equity Financing)

Corporate financing in energy infrastructure development is typically used for small scale domestic projects. For instance, this applies to those transmission or distribution projects that are within one TSO's service area. This type of finance primarily aims to maximise shareholder value. Shareholder loans are a debt-like process where financing is provided by equity. These loans' maturity dates are often long with minimum interest payments. The decision on what type of financing the company will apply to a certain project depends on many factors. The factors to be taken into account include the balance sheet of the project, its credit-worthiness, and ability to leverage debt and risks. Typically, in corporate financing the funds provided are in proportion to the shareholders' respective shares. These shares may include cash flows or debts issued against the company's assets. In this type of financing, the parent company's balance sheet will define the bankability of the project.

Energy infrastructure project financing imposes additional risks that may not be the case for instance in the construction sector. As explained below, for bigger infrastructure projects, such as LNG and pipelines, project financing is often the preferred method. Project financing could be an attractive option for smaller companies who are not willing to invest all their assets in one project. This said, the Caspian Pipeline Consortium,[33]

---

[33]Press release by Caspian Pipeline Consortium Director, Ian MacDonald, CPC: History, Reality and Future, 2005, available at: http://www.investkz.com/en/journals/43/23.html.

West African Gas Pipeline[34] and Baku-Supsa pipeline projects have all used shareholder loans. This was possible because these companies had an operating history, and in the case of the West African Gas Pipeline, only the Government of Ghana received external debt-financing to support its equity contribution.[35] Project sponsors may have a mix of equity finance and debt-funding from commercial banks, export credit agencies and/or multilateral financial institutions. The financial arrangements may become more complicated if there were multiple actors in the value chain, such as different owners of upstream facilities, transport and downstream assets, which is an inevitable result of liberalisation. This is not to criticise the unbundling per se. Without a doubt opening up competition offers substantial benefits to the end user. But these are just facts about the principles of financing multi-state, multi-year mega projects that are presented here to explain some of the reasons why public funding have been necessary in the past in an increasingly liberalised EU natural gas market.

4.3.2 Project Financing

As explained above, there are various financing methods which can be employed in order to develop energy projects. The most suitable among them for major projects, such as cross-border oil and gas pipelines, is project finance, which is also referred to as debt-financing. Project financing provides funds for a specific project company. Investors set up a project-specific company referred to as a SPV, to build, own and operate the energy infrastructure, financed by a mixture of equity and debt, often obtained from International Finance Institutions and commercial banks. The SPV offers no or limited recourse to the assets of the parent companies after the construction and operational stages of the project has commenced. Prior to the completion of construction, however, the shareholders may have to provide financial guarantees to the lenders.

---

[34]More information on West Africa Gas Pipeline Project is available at: http://documents.world-bank.org/curated/en/353691468767744635/West-Africa-Gas-Pipeline-Project.

[35]More information on Baku-Supsa, also known as the Western Route Export Pipeline is available at: https://www.bp.com/en_ge/bp-georgia/about-bp/bp-in-georgia/western-route-export-pipeline--wrep-.html.

Commonly, in major gas pipeline projects the debt ratio would be between 50 and 75%. The return of the investment, for equity share and the debt, is generated from the revenue stream of the project. For instance, the BTC pipeline and the North Stream projects (discussed below) have a 70/30 debt/equity ratio,[36] whereas the Chad-Cameroon pipeline had a 17/83 debt/equity ratio.[37]

In project finance, there has to be sufficient confidence from lenders that the sponsors have the technical capacity to complete and run the project over its lifetime. In the wake of the financial crisis, banks have been reluctant to provide funds for projects that have long payback periods, such as cross-border pipelines.[38] Commercial banks are similarly hesitant to support projects developed in 'new regulatory frameworks'. The Third Energy Package may not be considered new as it dates back to 2009. However, adoption and application of Network Codes, in particular on new and incremental capacity allocation, was tested for the first time in April 2017. In addition to the uncertainty with the implementation of new rules, the default regulated approach of the Third Energy Package posed some challenges for project financing.

In 2011, it was considered that following the unbundling and privatisation of the TSOs, additional investments were necessary to achieve the EU's 2020 energy security ambitions.[39] However, as explained above, all of the cross-border pipelines, the majority of LNG facilities and storage facilities have been constructed with public funds, instead of on the basis of purely commercial projects. These projects have been subject to NRA approved tariffs. In many EU Member States, the tariffs are cost-based in the form of a cost-plus or rate-of-return format. The rate of this standard regulated return, which varies across different

---

[36]More information on the investment structure of North Stream pipeline is available at: https://ppi.worldbank.org/snapshots/project/nord-stream-gas-pipeline-phases-i-and-ii-5867.

[37]More information on Chad-Cameroon pipeline's financial structure is available at: http://web.worldbank.org/archive/website01210/WEB/0__MENU.P.HTM.

[38]International Energy Agency (IEA), Word Investment Outlook, available at: http://www.iea.org/publications/freepublications/publication/weio2014.pdf, 35–37.

[39]The structuring and financing of energy infrastructure projects, financing gaps and recommendations regarding the new TEN-E financial instrument (European Commission Directorate-General for Energy, Berlin/Brussels, 31.07.2011).

Member States (as some allow a bonus system),[40] has been considered
a barrier to the viability of private infrastructure investment as the pay-
back period of debts obtained via project financing becomes fluid.

Project financing requires detailed due diligence studies and risk
assessment. Major energy infrastructure projects take a long time to be
realised, and they are further prolonged under the Third Energy Package
due to uncertainty with authorisation and permitting procedures,
which are implemented differently in each Member State. Some law-
yers involved in development of these projects have noted that this may
amplify the unwillingness of lenders to be involved in project finance.[41]
Furthermore, banks do not like volatility of cash flows because, after
the operations have started, lenders have limited recourse to the shareholders. Particularly in LNG projects, banks could be very wary of any
risk factors which may impact cash flows.[42] Lenders would like to see
security of supply and gas transportation throughout the lifetime of the
project alongside regulatory certainty. However, many Member States
do not necessarily offer a regulatory environment that adequately
favours conditions for gas infrastructure investments.[43] Hence, public support for natural gas project has been necessary to date (and it is
uncertain that this will continue after 2020, when the CEF funding
period comes to an end).

Despite the fact that project finance adds more costs to the project
than corporate finance (because of the cost of due diligence, insurance
covers, charges by lenders on the advanced funds and the costs of monitoring performances), project sponsors still seek it for large-scale energy
infrastructure projects in order to benefit from the advantages provided

---

[40]European Commission, study on regulatory incentives for investments in electricity and gas
infrastructureprojects—Final Report, 2014, available at: https://ec.europa.eu/energy/sites/ener/
files/documents/MJ0614081ENN_002.pdf

[41]René H. Gonne and Wim Vandenberghe, Investing in European Regulated Gas Infrastructure:
A Favourable Legal Environment? in *The International Comparative Legal Guide to: Gas Regulation
2012, A Practical Cross-border Insight into Gas Regulation Work*, Published by Global Legal Group,
in association with Ashurst LLP.

[42]Sophia Ruester, Financing LNG Projects and the Role of Long-Term Sales-and-Purchase
Agreements, Deutsches Institut für Wirtschaftsforschung, 2015.

[43]TYNDP 2015, 32.

by it. Project finance provides the long-term finance that is required by a project that has a high capital cost, which cannot be recovered within a short term without significantly increasing the final price of the project products. Pipeline investment with an estimated 20 years life-span will benefit from the high leverage despite the slow return of investment. Furthermore, the tax benefit is an additional advantage because the interest charged by lenders is tax deductible. It also affords the project sponsors an off-balance sheet finance allowing them to increase their credit rating and borrowing capacity. It is employed as a risk control technique since the liabilities of the investors are limited to the amount of their equity investments.

Overall, project finance affords the sponsors the opportunity of risk spreading and joint ventures even with unequal partners. The attraction of project finance also lies in the fact that sponsors could, to some extent, transfer some political risks to the lenders. Interest rates required by the lenders, however, tend to be higher than in a corporate financing framework. Below are some examples where a syndicate of commercial financing mechanisms was used to fund pipeline projects.

4.3.2.1 Example 1: North Stream Pipeline Project

The submarine section of the North Stream 1 pipeline project was funded by project finance. The project raised €3.9 billion in 2010 with financial backing from 26 banks. 30% of the investment was made through shareholder loans. Shareholders included: Gazprom (51%), BASF/Wintershall Holding GmbH (20%), E.ON Ruhrgas AG (20%), and N.V. Nederlandse Gasunie (9%). The Export Credit Agency (ECA) cover for the financing are HERMES and SACE with extra from the German Government's untied loan guarantee scheme. A further €800 million were covered by a commercial, syndicated local facility.[44] Among the reasons why North Stream 1 proved commercial viability was the transportation agreement which included ship-or-pay clauses.

---

[44]Press release, Nord Stream, Nord Stream Completes Phase I Financing, 16.03.2016, available at: https://www.nord-stream.com/press-info/press-releases/nord-stream-completes-phase-i-financing-341/.

North Stream 2 is being constructed under long-term financing at the time of writing in 2019. 50% of the fund is provided by European energy companies under an agreement signed in April 2017. These companies are: ENGIE, OMV, Shell, Uniper and Wintershall. Each company will provide loans covering 10% of the total costs[45] that are expected to reach €9.5 billion. Hence, each company will be funding a maximum of €950 million,[46] while the remaining 50% will be funded by Gazprom. In 2018, construction of the Nord Stream 2 gas pipeline was launched, with the pipeline expected to be put into operation before late 2019, doubling the current transportation capacity to 110 bcm per year. Project financing was not utilised as international sanctions on Russia make it difficult for International Finance Institutions to lend Russian corporations. Weak gas prices, demand uncertainty persisting at the time of writing in Europe, and high capex, could make it impossible to fund the project for Gazprom alone on the basis of equity finance.

4.3.2.2 Example 2: Turk Stream

The Turk Stream project's price tag is expected to reach €11.4 billion, with the first line costing approximately €4.3 billion. The offshore segment could cost around €7 billion; Gazprom will provide €310 million to the project company, South Stream Transport (also a subsidiary of Gazprom), to develop it.[47] The economic gains of the Russian state budget from export taxes are expected to reach US$750 million. For Turkey, the economic advantage is a discount in the price of gas and, in the event the

---

[45]News Article, *Financial Times*, Gazprom to Receive Funding for Nord Stream 2 Pipeline: Five European Energy Groups Will Provide Loans to Contentious Project, available at: https://www.ft.com/content/32898bae-28f3-11e7-9ec8-168383da43b7.

[46]Press release, Wintershall, Nord Stream 2 AG and European energy companies sign financing agreements, 24.04.2017, available at: https://www.wintershall.com/press-media/press-releases/detail/nord-stream-2-ag-and-european-energy-companies-sign-financing-agreements.html.

[47]Gokce Mete, Turk Stream Pipeline Project: An Analysis of Legal, Financial and Technical Aspects. European Centre for Energy and Resource Security, Reflections, Working Paper Series Volume 3, Spring 2017, available at: https://www.kcl.ac.uk/sspp/departments/warstudies/research/groups/eucers/pubs/reflections-3.pdf.

second line is built, the profits from transit fees. Finally, should it be able to resell the gas, Turkey will receive any subsequent profit thereof. The offshore section of both the first and second lines will be fully owned and financed by Gazprom. BOTAŞ, on the other hand, will create a company for the construction of the first onshore 180 kilometre-long line which will supply gas to the Turkish market. For the second onshore line, a 50-50 joint venture company between Gazprom and BOTAŞ's subsidiary will be established. The costs of the second onshore line will be paid by the participants of the joint venture, proportionate to their respective shares. If both sides agree, third party funding could be sought for the second onshore line and a third participant could be added to the joint venture.

For the second line, Italy and Greece are likely destinations for the Russian gas via the Southern Corridor, if Russian transit through Ukraine partly decreases after 2019; and these volumes are redirected to North Stream 2 via Germany. The price of gas will likely increase when it reaches Italy, in particular due to additional delivery tariffs and supplementary costs. If TAPs current capacity of 10 bcm is expanded to 20 bcm, by adding two compressors in Greece and Albania, Russia can use this capacity via Third-Party Access. By transporting larger volumes (of both Russian and Azeri gas) to Europe the overall network costs for TAP operators would be reduced.[48]

4.3.2.3 Example 3: Baku-Tbilisi-Erzurum Oil and South Caucasus Gas Pipelines

The BTC pipeline is owned, developed and operated by the integrated Baku Tbilisi Ceyhan Pipeline Company ("BTC Co."). BTC Co.'s shareholders are as follows: BP (30.1%); AzBTC (25%), Chevron (8.90%), Statoil (8.71%), TPAO (6.53%), ENI (5.00%), Total (5.00%), ITOCHU (3.40%), INPEX (2.50%), ConocoPhillips (2.50%) and

---

[48]Gokce Mete, Turk Stream Pipeline Project: An Analysis of Legal, Financial and Technical Aspects Publication, European Centre for Energy and Resource Security, Reflections, Working Paper Series Volume 3, Spring 2017, available at: https://www.kcl.ac.uk/sspp/departments/warstudies/research/groups/eucers/pubs/reflections-3.pdf.

ONGC (BTC) Limited (2.36%).[49] In Turkey, BOTAS, the state-owned crude oil and natural gas pipelines and trading company, entered into a lump-sum turnkey contract and constructed the Turkish section of the pipeline. Following a similar structure, the SCP pipeline is owned and operated by the integrated SCP Company ("SCP Co.").

The shareholders include BP (28.8%) as the operator, AzSCP (10.0%), SGC Midstream (6.7%), Statoil (15.5%), Lukoil (10%), NICO (10%) and TPAO (19%).[50] SCP Co. was responsible for the construction of the pipeline sections in both Azerbaijan and Georgia. Then, on behalf of the BP-led consortium, BOTAS constructed the pipeline infrastructure from the Turkish-Georgian border to the Turkgozu entry point at Erzurum. While the target for the BTC pipeline is western markets, the SCP was built to export gas to Turkey (6.3 bcm a year). Despite this, sales contracts for gas produced from the Shah Deniz field Phase I were also signed with Azerbaijan (1.5 bcm a year plus 3 bcm in 2006–2008) and Georgia (0.8 bcm a year).[51] The BTC Project was largely financed by the EBRD, IFC, and a syndicate of 15 commercial banks.[52] The oil revenues generated by successful operation of the BTC pipeline aided Azerbaijan and the Shah Deniz consortium companies to finance the SCP pipeline, with further support by a loan of approximately US$100 million to Shah Deniz and US$60 million to South Caucasus, provided by EBRD on behalf of SOCAR. EBRD's syndicated loan security included completion guarantees by BOTAS (a lump sum turnkey contract was signed for the Turkish section of the pipeline backed by government guarantees), share pledges both from the Shah Deniz consortium and SCP, assignment of SCP Co.

---

[49]See business updates of BP Caspian, available at: http://www.bp.com/en_az/caspian/press/businessupdates/businessupdates.html.

[50]See business updates of BP Caspian, available at: http://www.bp.com/en_az/caspian/press/businessupdates/businessupdates.html.

[51]News Article, Trend Az, Shah Deniz 2 Start-up Is Milestone for Entire Caspian Region—Wood Mackenzie, 02.07.2018, available at: https://en.trend.az/business/energy/2923820.html.

[52]See BP in Georgia at: http://www.bpgeorgia.ge/go/doc/1339/150562/Baku-Tbilisi-Ceyhan-BTC-Pipeline-.

and Lukoil Midstream's rights under project agreements, and pledges from SCP Co's and Lukoil Midstream's offshore bank accounts.

The approximate financing arrangement of BTC included IFC, EBRD and AB loans worth US$500 million, over US$700 million loan given by ECA facilities, with the Japan Bank for International Cooperation Overseas offering a US$300 million investment loan, Overseas Investment Corporation gave US$100 million loan and a further US$900 million in sponsor senior loans were included in the debt which covered up to a 12-year term. A further US$1 billion came from 15 commercial banks. The success of being eligible for project financing could be explained, in part, by the fact that Shah Deniz had adequate reserves to fill the pipelines and there was substantial financial commitment from project sponsors in terms of equity. The project also proved capable of guaranteeing a minimum rate of return regardless of volumes.[53]

Tariffs are payments received by project participants (there may be many project participants) from shippers and other customers for the transportation of petroleum through the pipeline system. In practice, tariffs may be calculated under different methodologies, such as point-to-point, or entry/exit or distance-based.[54] They may also be structured in accordance with domestic laws for each section of the pipeline or determined on a contractual basis. In the projects in question, it appears that the latter structure was preferred. Hence, the tariff methodology is not regulated under the Host Governmental Agreement (HGAs) but left to be negotiated under commercial contracts between project investors (there may be many investors and pipeline owners) and potential shippers. Nevertheless, the SCP and Azerbaijani BTC project agreements provide guidance on the methodology for the allocation of tariffs, which are based on a reasonable method taking into account the

---

[53]Chadbourne and Parke Rubin Weston, The Project Financing of Cross-Border Pipelines (Energy Charter Workshop, Brussels,17 October 2006).

[54]Energy Charter Secretariat, Gas Transit Tariffs in selected Energy Charter Treaty Countries (2006).

length of the pipeline in the territory.[55] In respect of tariff allocation, an additional criterion to consider is the amount of capital expenditures or expected capital expenditures incurred, as was the case in the Azerbaijani HGA of the BTC.[56]

Transit fees in the BTC HGAs are defined as taxes levied on revenues generated from the petroleum transported through the capacity owned by each shareholder.[57] The transit fee paid to Turkey is the corporate tax, calculated on the basis of the *"taxable income of the project participants related to project activities"*.[58] In Georgia, the transit fee is described as a profit tax, the calculation of which is based on the project participants' *"separate share of any income, expenses and other taxable items related to project activities for the year, as determined in accordance with the Tax Code of Georgia and deductions"*.[59]

In the SCP project, the transit fees to be paid to Georgia are determined as a combination of taxes and payments in kind, that is, an offtake of gas from the pipeline for its domestic use. Accordingly, Georgia may dispatch 5% of the gas throughput or, instead, opt to collect transit fees in the form of a 'minimum tax amount' fixed at US$2.5 per million cubic metres (mcm) for the first year, subject to a 2% increase per year.[60] SCP Co. is also liable for profit tax (identically calculated as that in the BTC Georgia HGA). However, while determining its taxable income and loss, such minimum tax incurred will be deducted.[61]

4.3.2.4 Observations

The Nord Stream and BTC pipeline projects demonstrate that for an international pipeline project to be bankable, it requires a robust upstream production or development with strong capital (suppliers),

---

[55]Art. 8.2 (v) SCP HGA.
[56]Art. 8.2 (v) Azerbaijan BTC HGA.
[57]Art. 9.2 (iii) Turkey BTC HGA.
[58]Art. 9, Turkey BTC HGA.
[59]Art. 8, Georgia BTC HGA.
[60]Art. 8.3, SCP HGA and Appendix 1.
[61]Art. 8.2 (i) and (vii) 15.

the existence of downstream markets and creditworthy suppliers (buyers). The bankability of these projects also required a sound contractual framework that addresses commercial risks throughout the entire value chain, an enforceable legal regime for contracts and intergovernmental agreements and cash-flow guarantees via ship or pay obligations (to avoid volume and market risk).

With the principles of project financing in mind, many have criticised the liberalisation of the markets since 1998 in the EU. Critics contemplated that liberalisation led to instability in terms of financing large scale projects—as the short markets created do not support financing mechanisms of capital intensive energy infrastructure projects. This criticism is not justified: the risks to financeability of these projects from shorter markets, unbundling and third party access may be addressed under possible alternatives available to EU policymakers. Next, a discussion on how best to maintain and attract private sector investments in natural gas projects in order to avoid carbon locking with public funds will be provided. These investments will still be necessary, as explained in the previous chapters, in particular in Central and South East Europe where the internal gas market is not yet complete, and in security of supply overall, hence import infrastructure, to replace declining domestic gas production. Nonetheless, these alternatives have some pros and cons to consider, and the availability or lack thereof of funds can have a sizeable impact on the future of gas in the EU.

4.3.2.5 Example 4: Chad-Cameroon

Another example is the Chad-Cameroon oil pipeline project.[62] Despite the absence of adept governance organisations, legal mechanisms and environmental regulations to stabilise the project in both countries, it became operational in 2004, as a result of the collaborative efforts of the IFC and the World Bank (WB).

---

[62]Factual information regarding the Chad-Cameroon pipeline project can be found at: http://web.worldbank.org/.

In this project Chevron, Exxon and Petronas funded 80% of the project's costs amounting to US$1.3 billion. The IFC and ECAs, together with 18 commercial banks, provided 17% of the projects cost (US$600 million). The WB and EIB's support corresponded to 3% of the projects cost (US$135 million). A US$400 million bond was issued by the project sponsors but later cancelled due to lack of interest by the market. The funding gap was filled by equity financing.

The funding that comes from multilateral agencies (IFC, EBRD, ADB, etc.) can facilitate the decision by commercial banks to participate in projects. The IFC and WBG agreed to finance the project, as they were convinced that the outcome of the project would remove poverty through management of the pipeline's revenues and promote reforms leading the countries' development. The WB conditioned the final approval for its loan to the respective Chadian and Cameroonian Governments upon compliance with a comprehensive set of assurances to ensure that the revenues would be used to cut poverty and improve living standards, and that environmental safeguard mechanisms would be fulfilled. To achieve this, the WB required Chad to deposit all royalties, taxes and dividends accrued from the project to an escrow account. Further, WB conditions included the requirement that 10% of these revenues were to be allocated for future generations, 80% to the welfare of citizens in terms of education, health, water management, social services and rural development, 5% for upstream development and 15% to the state treasury. As these terms were subsequently breached by the Chad Government in 2008, the WB fund has had to be repaid.

4.4   Non-traditional Financing Methods

There is a growing trend by non-traditional funds such as pension funds, insurance companies and wealth funds to invest in infrastructure assets. TSOs in the EU struggle to attract large private funds into energy infrastructure investments. This funding gap has been filled with public subsidies to date. These non-traditional financing methods could respond to the urgent need for finance and equity to support European energy integration. For the investors, infrastructure assets offer diversification opportunities for their assets and a long investment horizon with steady cash flows.

Private sector investments are increasingly scarce, and investors are reluctant to invest in infrastructure that will take a long time to gain commercial return. This is the reason that no gas pipelines selected as PCI were constructed without public subsidies. However, public sector finance is also restricted, as many national governments have still not overcome the effects of the economic crisis of 2008. Furthermore, the EU will need to invest heavily in new low carbon technologies, including renewables, battery solutions, CCUS and in low carbon and green gases in order to meet its 2050 net zero carbon goals. To achieve the objectives of the Energy Union, the Clean Energy Package and the Third Energy Package, new methods of financing will have to be explored, especially in order to support new infrastructure and interconnectors.

Trust funds and pension plans have been used to finance upstream and midstream asset development in the US, as an alternative to private debt, equity funds, or the use of public resources (through master limited partnerships). Furthermore, real estate investment trusts have also been used to raise capital to provide debt financing for the upstream and midstream oil and gas sectors in the US. In Europe, in recent years, there has been growing interest from specialist infrastructure funds, pension funds and private equity firms in the North Sea midstream sector which generates returns on investment via transmission tariffs under long-term contracts, therefore offering stable long-term revenues. In 2016, a US private equity investor purchased Total's 100% interests in both the Frigg UK pipeline and St. Fergus gas terminal, as well as a 67% interest in the Shetland Island Regional Gas Export System pipeline.[63] As the EU market has leaned towards short-term contracts (less than one year) the same could not be said in terms of investment

---

[63]Press release, Total, UK: Total Sells North Sea Midstream Assets for £585 Million, 27.08.2015, available at: https://www.total.com/en/media/news/press-releases/uk-total-sells-north-sea-midstream-assets-ps585-million.

attraction.[64] Furthermore, public opposition towards fossil fuel investments of pension schemes is forcing EU governments to disinvest shares
of pension schemes in oil and gas projects.[65] The UK and Sweden are
among the latest national funds to sell their shares in oil companies that
operate in breach of the Paris climate agreement.[66]

In addition to this, resource-rich countries like Norway, Israel, the
US, Kuwait and Chile have created sovereign wealth funds as permanent trust funds accumulating oil and gas revenues. This has helped
these countries with new sources of capital for investments, while at
the same time assisting them in managing commodity price cycles.
A consumer version of these sovereign wealth funds in the EU is the
European Fund for Strategic Investments (EFSI). Creating seed funding
to mobilise private investment in projects which are strategically important for the EU is a preferred solution. In an ideal world, these resources
should be allocated for commercially viable projects as the EFSI fund is
essentially a loan that will need to be paid back. This is preferable to the
CEF type subsidies which can be used to subsidise political/emergency
infrastructure rather than commercial projects. Financing investments
in oil and gas projects are prone to changes in policy and politics which
may create mismatches for linking funding sources and opportunities
for investment.[67]

In addition to pension funds, insurance companies and wealth funds,
the role of trades in financing infrastructure should also be considered.

---

[64]White and Case Publication, Investment in Oil and Gas Midstream Infrastructure,
28.10.2016, available at: https://www.whitecase.com/publications/alert/investment-oil-and-gas-
midstream-infrastructure.

[65]News Article, *The Guardian*, and Boost for fossil fuel divestment as UK eases pension rules,
Exclusive: pension schemes will be free to dump fossil fuel investments after government
drops, best returns, legal rule, and 18.12.2017, available at: https://www.theguardian.com/
environment/2017/dec/18/boost-for-fossil-fuel-divestment-as-uk-eases-pension-rules.

[66]News Article, Reuters, Swedish Pension Fund Sells out of Six Firms It Says Breach Paris
Climate Deal, 15.06.2017, available at: https://www.reuters.com/article/us-climatechange-in-
vestment-sweden/swedish-pension-fund-sells-out-of-six-firms-it-says-breach-paris-climate-deal-
idUSKBN1962CC.

[67]John Mitchell, Valérie Marcel, and Beth Mitchell, Oil and Gas Mismatches: Finance,
Investment and Climate Policy, Research Paper, Energy, Environment and Resources, Chatham
House, July 2015, available at: https://www.chathamhouse.org/sites/files/chathamhouse/field/
field_document/20150709OilGasMismatchesMitchellMarcelMitchellUpdate.pdf.

They are increasingly willing to fund these deals as they have access to huge lines of credit at low rates. These funds tend to have an ability to better identify investment opportunities and they are ready to invest in debt or equity instruments especially when this enhances their trading business.

The lack of private investment in natural gas networks in the EU is not believed to be due to lack of capital. Since there are enough funds chasing capital, the problem is the lack of clarity over rate of return in energy infrastructure projects. Several studies conclude that there will be a significant increase in the allocation of pension funds into the infrastructure sector, compared to other sector investments such as property. For instance, a survey by First found that 73% of senior pension fund professionals say they expect schemes to increase their allocation to infrastructure. Some of the world's largest pension funds also show specific interest in green energy projects.[68] They need to know how much they will earn from the investment as they are often wary of greenfield projects that usually carry larger risks. Some suggest that if the government underwrites some of the risks with large energy projects, then pension schemes would be more willing to fund them.[69] Pension funds also find the administrative burden of direct investment in infrastructure projects onerous. Thus, there is a tendency to acquire projects that are already constructed and in operation.[70]

Therefore, considering the limited availability of trust and pension type funding, the final investment decision and project design should ideally be based on commercial realities. Decisions should focus on the effective functioning of free and open markets, which would inevitably mean that the most competitive sources of commodities will enter the market, with the cheapest price tag, which would ultimately benefit end users. This would allow public funds to be freed up for financing more innovative and cleaner energy technologies of the future.

---

[68]OECD (2011), Pension Funds Investment in Infrastructure: A Survey, available at: http://www.oecd.org/futures/infrastructureto2030/48634596.pdf.

[69]News Article, *Financial Times*, Pension funds crave more infrastructure projects, 21.10.2016, available at: https://www.ft.com/content/a05fe960-95ec-11e6-a1dc-bdf38d484582.

[70]News Article, *Financial Times*, Pension funds crave more infrastructure projects, 21.10.2016, available at: https://www.ft.com/content/a05fe960-95ec-11e6-a1dc-bdf38d484582.

4.5   Sustainable Finance Package

Another reason why, despite their suitability for financing large infrastructure projects, pension funds and trust funds may not be able to offer a response to the problem of raising financing in the EU for natural gas pipeline projects is because on 24th May 2018, the Commission released a set of legislative proposals on financing sustainable growth (the so called Sustainable Finance Package).[71] This Package includes proposals to establish a unified sustainable finance 'taxonomy', in other words, a common classification system and to introduce new environmental, social and governance duties and disclosures. The EU Taxonomy is essentially an implementation tool that can enable capital markets to identify and respond to investment opportunities that contribute to environmental policy objectives. It aims to provide a non-exhaustive list of economic activities assessed and classified based on their contribution to EU sustainability-related policy objectives and particular to climate change mitigation.

In order to implement this action plan for financing sustainable growth, the Commission established a Technical Expert Group (TEG) on sustainable finance in July 2018.[72] TEG published a technical report on EU taxonomy on June 2019 which contains a technical screening criteria for 67 activities across 8 sectors that can make a substantial contribution to climate change mitigation and a methodology for evaluating substantial contribution to climate change adaptation, which includes the energy sector. The report acknowledges that CCS facilitates the direct mitigation of both fossil and process emissions in many industrial sectors including steel, cement and chemicals. For activities which go beyond 2050, unabated natural gas-fired power generation is not expected to meet the required threshold, however gas-fired power with CCS may qualify. TEG has a mandate until the

---

[71]Proposal for a Regulation of the European Parliament and of the Council on the establishment of a framework to facilitate sustainable investment, COM (2018) 353 final, 24.5.2018.

[72]Technical Expert Group (TEG), Taxonomy Technical Report, 18.06.2019, available at: https://ec.europa.eu/info/sites/info/files/business_economy_euro/banking_and_finance/documents/190618-sustainable-finance-teg-report-taxonomy_en.pdf.

end of 2020 and its recommendations support the Commission in the development of the proposed taxonomy regulation.

The Sustainable Finance Package also includes proposals to create low-carbon and positive-carbon impact benchmarks on companies' carbon footprints. In February 2019, the EU Parliament and the Council reached a political agreement on the creation of two new categories of benchmarks: the 'EU climate transition benchmark' and the 'EU Paris-aligned benchmark'. The purpose is to provide harmonised indices to enable potential investors to evaluate the carbon footprint generated by the projects or assets they consider, before their decision to invest. The EU climate transition benchmark aims to lower the carbon footprint of a standard investment portfolio, and the EU Paris-aligned benchmark have the more ambitious goal to select only components that contribute to attaining the 2 °C reduction set out in the Paris climate agreement. Following EU ambassadors endorsement, the text is expected to be adopted by the Parliament and the Council as a next step.

Furthermore, a proposed regulation to introduce disclosure obligations on how institutional investors and asset managers integrate environmental, social and governance (ESG) factors in their risk processes, was also included in this action plan. And, in April 2019, the European Parliament endorsed legislation setting the building blocks of a capital markets union, including regulation on disclosures relating to sustainable investments and sustainability risks, which aims to strengthen and improve the disclosure of 'green' information by manufacturers of financial products and financial advisors towards end-investors.[73]

In brief, the proposed regulation and the work carried out so far by TEG is telling investors to choose green products. This package can therefore play a significant role in shaping business models and investments decisions in the energy sector in the decades to follow. It really has the potential to act as a catalyst for the EU to truly mobilise the financial sector to help fund the transition to a greener, more

---

[73]Press release, European Commission, Capital Markets Union: European Parliament Backs Key Measures to Boost Jobs and Growth, 18.04.2019, available at: https://europa.eu/rapid/press-release_IP-19-2130_en.htm?locale=en.

sustainable economy. In the interim, investors might adopt a wait-and-see approach, instead of massively investing into new low-carbon alternatives, as they know that individual policies can be quickly reversed. However, future adjustments to the legal framework on finance might make it more challenging for the oil and gas industry to obtain loans and grants from investment entities, as inclusion of a sustainable index or sustainable fund in an investment may crucially impact the prices of shares and the cost of debt funding.

It must be noted here that, while unlocking finance is of course essential to reach the objectives of the Paris Agreement, investment decisions will also depend on the cost and availability of technologies, societal impact of the energy transition and industrial competitiveness of the EU.

4.6   Commercial and Political Risk Mitigation

There are various other instruments to mitigate commercial and political investment risks in large energy projects. For instance, long-term contracts covering subscribed capacity allow for cost-recovery on contract duration. Regulatory certainty from the host states, which guarantee that conditions will not change during the contract term, could minimise political and commercial risk factors. If tariffs are known in advance, and if the TSOs receive fair remuneration, while taking into account the specifics of each project, risks can be controlled more effectively.

There are further international instruments available to help mitigate political risk in energy infrastructure projects. For instance, guarantees by the International Development Association (IDA).[74] For instance, the Risk Mitigation Facility (RMF) of IDA seeks to catalyse private sector investment in large-scale infrastructure and Public-Private Partnerships (PPPs) by providing the following risk mitigating products: (i) liquidity support instruments backstopping payment obligations

---

[74]More information is available at: http://ida.worldbank.org/.

of state-owned enterprises (SOEs) to private projects; and (ii) political risk insurance (PRI) and government counterparty coverage for project finance loans and equity investments. IDA guarantees would potentially cover commercial banks' lending to the pipeline project company for up to 100% of debt service defaults caused by: government non-performance, political risk, foreign currency convertibility and transferability risk, asset expropriation and changes in law. Only few IDA guarantees have been issued so far, but there is a rise in use of this method of risk mitigation.

Other mechanisms include Multilateral Investment Guarantee Agency (MIGA)[75] guarantees with the multilaterals who bring risk mitigation and credit support. MIGA is the political risk insurance arm of the WB. Prior to issuing MIGA guarantees, a due diligence process is carried out. MIGA political risk guarantees are well-suited to reduce non-commercial oil and gas investment risks. They are designed to help companies overcome hurdles that may loom large prior to deal signing, particularly for costly investments in countries seen as high risk. The assessment includes a requirement of the project to comply with a broad and comprehensive set of environmental and social performance standards, over and above the basic requirement of compliance with all applicable laws and regulations. The due diligence results in classification of the project depending on its impact on society and the environment. The results are made available to the public, and non-governmental organisations (NGOs) can respond to the public disclosure. However, the downside is that the entire process is quite lengthy and could delay the implementation of the project significantly.

MIGA offers political risk insurance solutions to protect investors in different scenarios. For instance, the inability of investors to convert or transfer dividends or loan payments due to foreign exchange restrictions. Another scenario is direct expropriation or creeping expropriation, where a host state makes it impossible to operate the project by adopting discriminatory measures. MIGA also offers coverage in case of war and civil disturbance, as well as in the case of breach of

---

[75]More information is available at: https://www.miga.org/.

contract by the government or its state-owned entity. Some of these instances may trigger an investment-state arbitration, and MIGA would only pay once a final award is rendered. In the case of expropriation, MIGA covers the net book value of the insured investment, which may be difficult to calculate in case of creeping expropriation. The coverage includes foreign investments in the form of equity, shareholder loans, shareholder loan guarantees, and non-shareholder loans. All loans and loan guarantees, including those issued by the project shareholders, must have a minimum maturity of more than one year provided that MIGA determines that the project represents a long-term commitment by the investors.[76] MIGA, for example, covered the South African investors SASOL pipeline project in Mozambique and the Bolivia–Brazil natural gas pipeline project.[77] MIGA guarantees the aim to mitigate non-commercial risks and reduce the cost of the investment.

The abovementioned non-commercial and commercial mitigation mechanisms do not offer a silver bullet. Large energy infrastructure projects carry long-term risks at all stages of the value chain and throughout their life cycle. The complex structure of these projects, involving increasingly numerous stakeholders, requires a long-term strategy bringing certainty to the risks and risk management capabilities, reflecting mitigation mechanisms that responds to these often conflicting interests. In this direction, some experts note that these project management risks, if managed by the public sector, which may lack project management skills and resources, may imply that the appetite for providing funding for these projects will be significantly lower.[78]

---

[76]Political risk insurance tools of MIGA are explained here: https://www.miga.org/Pages/Resources/MIGA%20products.pdf.

[77]MIGA guarantees keep oil and gas projects on track are explained by Mamadou Barry, Head of Oil, Gas, Mining and Chemicals Sectors, MIGA, World Bank Group here: https://www.miga.org/documents/miga_oil_gas.pdf.

[78]Frank Beckers and Uwe Stegemann, A Risk-Management Approach to a Successful Infrastructure Project, Mckinsey, Report 2013, available at: https://www.mckinsey.com/industries/capital-projects-and-infrastructure/our-insights/a-risk-management-approach-to-a-successful-infrastructure-project.

4.7 Financing Alternatives

Cross-border energy infrastructure projects are inherently complex. There are many barriers to their implementation, from the investment decision to the final stage of decommissioning. The involvement of different jurisdictions, even within the EU, makes the administrative, legal, political and operational aspects more complicated. As a result of this complexity, the financing of cross-border energy infrastructure, in particular network-bound transport projects, tend to be challenging. In the past, mega projects which were developed with great optimism, such as Nabucco, were cancelled due to lack of finances. Despite the slight increase in oil and gas prices in 2018, a commodity price slump has been in place since 2015. This led to a trend towards short-term contracts with spot prices, moving away from take-or-pay contracts. Destination clauses and demand uncertainties because of climate policies have made it even more difficult to raise finance for such projects. It is becoming more difficult for the private sector to fund upstream and midstream projects, and often there is a need to receive loans from international finance institutions.[79]

The IEA's World Energy Investment Outlook 2014 estimated that over US$6 trillion investment was needed in upstream gas in the next two decades.[80] In the last three years, there was a sustained decline in global energy investment. This has changed in 2018 when energy investment remained at US$1.85 trillion. The largest growth rate came however from power, and most investments are taking place in the US, India and China, and not the EU. The IEA also noted in 2019 that a worldwide continued demand growth for natural gas (and oil) requires accelerated approval procedures for new conventional upstream projects.[81]

---

[79]International Energy Agency, Gas Security in Europe: Summary of the analysis and recommendations provided to the Group of Seven (G7) 2015–2016, available at: https://www.iea.org/media/topics/engagementworldwide/g7/Report.pdf.

[80]IEA, World Energy Investment Outlook 2014, available at: https://www.iea.org/publications/freepublications/publication/WEIO2014.pdf

[81]IEA, World Energy Investment Outlook 2019, available at: https://www.iea.org/wei2019/overview/.

Under all scenarios forecasted by the IEA, energy supply investment needs to rise globally, but major capital reallocation would be needed to meet sustainability goals. Availability of financial resources to add new capacities to close the supply gap as a result of rapidly decreasing domestic natural gas production is also vital for the EU. Ideally, these funds should be raised in the private sector based on market fundamentals. More significant investments are urgently needed to upgrade and innovate its energy industry, if the EU is to meet its 2050 goals. Since energy investment in the EU has declined by 7% over the past three years, according to the IEA (2019), where will the funds come from? All the PCI projects have been constructed with EU subsidies, but there are concerns over the efficiency of the CBA and PCI methodologies. With the new Commission taking the office in late 2019, there is a need to redesign the current EU infrastructure investment and its governance. How could private sector investment be instigated to avoid carbon lock-in with public money, allowing the private sector to take the market and stranded asset risks, while at the same time developing incentives for maintaining the value of the gas infrastructure by promoting and facilitating investment in low carbon and renewable gases to steadily decarbonise the gas grid towards 2050?

4.8   Market Driven or Driven by "Command and Control"?

The previous chapters, which looked at the EU natural gas regulation since 1990s, have already revealed that investments are not decided by the market per se. Ideally, and as a rule, market regulation is necessary to deal with market failures due to the lack of functioning wholesale markets (for instance in the case of competition law) and to address the abuse of market dominance, as energy transport infrastructure is a natural monopoly (such as third-party access and tariff regulation). Furthermore, regulation may be necessary for security of supply reasons.[82]

---

[82]Menno Van Benthem and Bert Tieben, The Optimal Regulatory Framework for the EU Gas Market a Discussion Paper on Regulation and Welfare Maximization SEO Amsterdam Economics, 10.05.2017, available at: http://www.seo.nl/uploads/media/DP_89_The_optimal_regulatory_framework_for_the_EU_gas_market.pdf.

In the EU, Kim Talus defines the regulatory timeline as a period of 'liberalisation' or 'deregulation' measures during 1990s and 2003 (the first and second energy packages) and the period after 2009 as 're-regulation'.[83] The re-regulation period, which is continuing well into 2019 presents more and more public-sector involvement that could be defined as a top-down approach to market intervention, rather than bottom-up. Essentially, the Third Energy Package is introduced to promote market-based mechanisms, but in reality, its implementation has resulted in an increased role for the EU institutions in commercial decision-making. Kim Talus defines this trend as a 'command and control' regime and criticises the EU energy market for being driven by political rather than commercial reasons.

Similarly, the cross-border pipeline projects used as case examples (some that are presented in the previous chapter, and more to follow in the next chapter) reveals that the Commission has great influence on deciding which cross-border import projects should be constructed. For instance, the South Stream pipeline project has been cancelled as a result of the Commission's intervention, and projects such as the Southern Corridor have been granted an exemption from the application of the Third Energy Package, again from the Commission. The OPAL pipeline exemption case displays a perfect example of the motivations being political in nature rather than commercial. Despite there being no shipper other than Gazprom to use the capacity of the pipeline, it has been a long battle to secure the remaining non-exempt capacity for the Russian state-owned company.

Cross-border pipeline projects that have proven commercial would not need to be subsidised. The current approach to market regulation is neither fully regulated nor an open market (where the market decides on security of supply).[84] Hence, an alternative recommendation for the

---

[83]Kim Talus, Decades of EU Energy Policy: Towards Politically Driven Markets, *The Journal of World Energy Law & Business*, Volume 10, Issue 5 (1 October 2017), pp. 380–388.

[84]Blog Post by Tim Boersma, What's Next for Europe's Natural Gas Market? 15.03.2016, available at: https://www.brookings.edu/blog/order-from-chaos/2016/03/15/whats-next-for-europes-natural-gas-market/.

Commission would be to adhere to the EU's liberal market ideology when ensuring the full implementation of the Third Energy Package, and throughout application of the Incremental Capacity Network Code, to ensure that the market is driven by commercial realities.

4.9   Invest in Upstream Assets in Third Countries

The financing options as alternative to traditional project finance and shareholder finance analysed in this book include public subsidies, public seed loans, pension and trust funds, and also soft methods like a change in policy to let the market take the risks. Another soft-method is facilitating EU companies' investments in resource rich, non-EU countries. There has been a decline of around 15% in worldwide investment in oil and gas pipelines since 2015, due to reduced activity in upstream oil and gas operations. The decline in investment in exploration has been explained by the oil price collapse since mid-2014, which raised concerns over a possible shortfall of resources to meet rising global demand.[85] The EU is the largest energy importer in the world.[86] Only around 15% of world's biggest oil and gas operation companies are European undertakings with international assets.[87]

It is true that a significant amount of the world's gas resources is located in areas with high security and geopolitical risks, such as Iran and Iraq. However, international oil companies have been operating in high risk territories for decades and, as explained below, in most cases, there are mechanisms of international law which can often mitigate these risks. Of the European companies operating in potential gas suppliers to the EU, the Italian Eni has been present in Turkmenistan since

---

[85]International Energy Agency (IEA), World Energy Investment Outlook 2017.

[86]IEA, Energy Policies of IEA Countries 2014 Review: European Union, Executive Summary, available at: https://www.iea.org/Textbase/npsum/EU2014SUM.pdf.

[87]Forbes List of the World's Biggest Public Companies, available at: https://www.forbes.com/global2000/list/#industry:Oil%20%26%20Gas%20Operations.

2008 in the Exploration and Production sector,[88] and the Norwegian company DNO entered Kurdistan in 2004, as one of the first international oil companies in the region, and has a leading position in reserves and production.[89] The French company Total is working with Russia in the development of Yamal LNG, a strategic liquefied natural gas project.[90] Total is in fact the second-largest private global LNG player in the world. In 2019, Total agreed to acquire Anadarko's assets in Algeria, Ghana, Mozambique and South Africa.[91] BP plc (British Petroleum) opened its Baku office in 1992 and has been operating the mega projects – Azeri-Chirag-Gunashli, Shah Deniz, BTC and the South Caucasus Pipeline with the Government of Azerbaijan since then.[92] BP's involvement in the Azeri Guneshli gas field has facilitated construction of the TANAP and TAP projects. More importantly, BP and its partners have signed gas sales contracts with European buyers, and although some of the Shah Deniz contracts are also oil-linked, others are priced off European gas trading hubs.[93]

The Commission and EU public institutions may like to consider rechannelling resources that are used for non-commercial and political midstream projects. Instead, they may like to consider providing seed financing for European oil and gas companies' upstream investments in countries where there is limited geopolitical and commercial risk in the upstream sector. This would also enable EU companies to influence

---

[88]More information on ENI's activities in Turkmenistan can be accessed at https://www.eni.com/en_IT/results.page?question=what+investments+does+eni+have+in+turkmenistan%3F.

[89]More information on DNO's operations in KRG is available at: http://www.dno.no/en/operations/where-we-operate/kurdistan1/.

[90]Press release, Total, Total Announces the Approval of the Development of Yamal LNG, a Strategic Liquefied Natural Gas Project, 18.12.2013, available at: https://www.total.com/en/media/news/press-releases/total-annouces-approval-development-yamal-lng-strategic-liquefied-natural-gas-project.

[91]Press release, Total, Total Agrees with Occidental to Contingent Acquisition of Anadarko's Assets in Africa, 05.05.2019, available at: https://www.total.com/en/media/news/press-releases/total-agrees-occidental-contingent-acquisition-anadarkos-assets-africa.

[92]More information on BP's operations in the Caspian region is available at: https://www.bp.com/en_az/caspian/aboutus.html.

[93]News Article, *Financial Times*, Azerbaijan Gas Pipeline Aims to Carve Out a Niche Across Europe, 01.01.2014, available at: http://www.ft.com/cms/s/0/174b403e-6c87-11e3-ad36-00144feabdc0.html.

investment decision-making in terms of supply markets in these countries and also contribute to sustainable development outcomes through corporate social responsibility practices which are necessary to obtain a social license to operate. For this to happen, the host country must be open to foreign investment, and there shall be legal guarantees against host states actions, such as expropriation without fair compensation. This was one of the reasons why the Energy Charter Treaty[94] was signed in the first place, to enable European companies to have access to upstream resources in oil and gas rich, former Soviet Union countries which were in a legal void in 1990s. The ECT is still effective, despite the biggest gas supplier of the EU, Russia, withdrawing from its provisional application in 2009. The ECT provides legal protection for foreign direct investment in the energy sector.

Another point of consideration is separating politics from the energy sector. The UN sanctions on Iran in the past have prevented European companies from developing large gas deposits in this country. In 2018, ongoing US sanctions, which are also supported by the EU Member States, resulted in ENI's suspension of its exploration activities in Russia's sector of the Black Sea which it was carrying out with state-controlled Rosneft.[95] It may be presumed that the EU's energy security would be improved with more influence from European businesses in the world's largest oil and gas operations. However, the volatility of oil and gas commodity prices, the Paris climate change accord, concerns over the EU's relationship with authoritarian regimes and human rights considerations are making it increasingly difficult for the EU to meaningfully engage with these countries and conduct business.

From the assessment presented in this book, it can be seen that the EU is committed to leading the decarbonisation agenda but at the same time openly promotes fossil fuel development, which is a sector that still receives considerable subsidies. Could an overarching reform in fossil fuel subsidies or a rigorous carbon price could offer solutions?

---

[94]2080 UNTS 95; 34 ILM 360 (1995).
[95]News Article, Natural Gas World, ENI Quits Russian Black Sea Project: Report, 30.03.2018, available at: https://www.naturalgasworld.com/eni-quits-russian-black-sea-project-report-60038.

4.10 Fossil Fuel Subsidy Reform

As mentioned above, despite the EU's pledge to tackle climate change, there is in fact no substantive reduction in oil, gas and coal subsidies. Fossil fuel subsidies take two forms; consumption subsidies and production subsidies. Consumption subsidies are intended to reduce the price of energy for end users. This can be executed by the government by introduction of a cap on petroleum, power or natural gas prices. Production subsidies are often used to reduce the cost of oil and gas resource extraction, and can be executed, among others, through tax incentives or government funding for fossil fuel production.

The European Commission reported that the UK in fact has the highest fossil fuel subsidies in the EU with around €12 billion a year. The level of such subsidies has not changed since 2008 levels, despite a EU commitment to phase them out.[96] The picture is not all that gloomy: while Spain, Germany and Italy all provide fossil fuel subsidies, remarkably they have provided more subsidies to the renewable energy sector. Yet, the UK is not alone, as the Netherlands, Sweden, France and Ireland gave more subsidies to fossil fuels than other energy sources.[97] The commitment to cut subsidies are not only made at the EU level but also among the G20 and G7.

Subsidising fossil fuels undermines climate change, energy security and sustainable development objectives. Fossil fuel subsidies also put a restraint on public budgets, limiting the funds that can be spent on economic diversification, education and health alike. They also distort competition with low carbon energy technologies. Therefore, fossil fuel subsidy reform is an important policy issue for all countries.

---

[96]Report from the Commission to the European Parliament, the Council, the European Economic and Social Committee and the Committee of the Regions, Energy Prices and Costs in Europe, Brussels, 9.1.2019, COM (2019) 1 final, available at: https://eur-lex.europa.eu/legal-content/EN/TXT/PDF/?uri=COM:2019:1:FIN&from=EN.

[97]Study on Energy Prices, Costs and Subsidies and their Impact on Industry and Households, Annexes to the Final Report, Trinomics, 03.09.2018, available at: https://ec.europa.eu/energy/sites/ener/files/documents/energy_prices_and_costs_-_final_report_-_annexes_v12.3.pdf.

The Commission acknowledges the problem and noted that reform of these structures is likely. However, they will need to be integrated into the country's broader aims for an energy transition. For instance, fiscal incentives for renewable energy projects such as feed-in tariffs and premium price or tax breaks might be more effective than abolishing these subsidies overnight and introducing a carbon price. The distributional impacts of green fiscal reforms have to be taken into account, and these changes should be ideally implemented gradually. Carefully crafted policies can also mitigate the negative impacts of removal of fossil fuels on poor households. It is also important that savings from fossil fuel subsidy reform are used to establish compensation schemes for the poorest groups of society, and that these savings can be used to mobilise infrastructure investments required in the energy sector, such as decarbonising hard to decarbonise sections of the economy.

4.10.1 Carbon Pricing

Carbon pricing could also have a significant impact on the pace of energy transition and the profitability of fossil fuel industries. Carbon pricing (along with removal of inefficient fossil fuels) makes low carbon solutions, such as renewables and CCS, more cost competitive. There are two main types of carbon pricing: emissions trading systems (ETS) and carbon taxes.

An ETS, also referred to as a cap-and-trade system—caps the total level of GHG emissions and allows those industries with low emissions to sell their extra allowances to larger emitters. By creating supply and demand for emissions allowances, an ETS establishes a market price for GHG emissions. The cap helps ensure that the required emission reductions will take place to keep the emitters (in aggregate) within their pre-allocated carbon budget.

A carbon tax directly sets a price on carbon by defining a tax rate on GHG emissions or—more commonly—on the carbon content of fossil fuels. It is different from an ETS in that the emission reduction outcome of a carbon tax is not pre-defined but the carbon price is. From an administrative perspective, a carbon tax is easier and faster to implement.

Putting a price on carbon is one of the most widespread policy options, as it is considered to be one of the most cost-effective way to drive reductions in GHG emissions and promote investments into alternative technologies. The Paris Agreement (Article 6) forms the legal basis for the establishment of market-based climate change mitigation mechanisms. The role of carbon pricing is likely to grow towards 2030. Already more than half of the signatories of the Paris Agreement referenced carbon pricing as a tool to achieve their national climate commitments in their Nationally Determined Contributions (NDCs). The choice of the instrument will depend on national and economic circumstances. There are also more indirect ways of more accurately pricing carbon, such as through fuel taxes, the removal of fossil fuel subsidies, and regulations that may incorporate a 'social cost of carbon'. GHG can also be priced through payments for emission reductions.

The EU ETS market, set up in 2005, is the world's first major carbon market and remains the biggest one. It operates currently among 31 countries (EU 28 plus Iceland, Liechtenstein and Norway), and covers around 45% of the EU's GHG emissions from power stations, energy-intensive industry sectors and airlines operating between these countries. In simple terms, it makes burning fossil fuels more expensive for these utilities and aims to incentive the use of lower emitting sources of energy. While the EU ETS suffered from too many or too few permits with a lack of balance in the past, it has resulted in coal and oil to natural gas switching as a transition or bridge fuel, particularly in Germany, Italy, Spain and the UK. The UK used a carbon price floor, which could be useful as a backstop measure to give long-term visibility and predictability to investors. Some countries in the EU topped the EU ETS with a carbon tax to cover sectors not covered by the ETS. Going forward, it may in fact be necessary to target areas that are not covered by the ETS. The EU ETS could also benefit from the creation of a uniform carbon price. However, this is not expected to take place before the 2030s.

In principle, the ETS was supposed to benefit the renewable energy sector, and other low carbon solutions. Instead, it appears that natural gas has been the main beneficiary due to lack of alternatives to replace natural gas in the heavy industries, transport and aviation sectors. Going forward, tighter carbon prices may encourage more investment in flexible and easy to store renewable and low carbon gases.

4.10.2 Next Cycle of CEF Funding for 2021–2027

As this chapter has shown, in principle it is possible to develop large-scale natural gas projects on a commercial basis, with some loans from EBRD, EIB, WB and commercial banks. Therefore, if the decision to invest is left to the investor, economically feasible projects with appropriate rate of return can work without the need for subsidies. This would also mean that competition would work and the cheapest gas and most efficient project(s) would get the go ahead. However, and this will be discussed further in the next chapter, in the EU natural gas infrastructure decision are not taken by the market but by political bodies. This has inevitably led to the need for public subsidies, and hence CEF and other fund structures were created. There are indeed mechanisms that can be utilised to avoid this scenario, but with the climate policies and a crusade against the natural gas sector under the Sustainable Finance Package, it will increasingly be difficult to even raise funds on a commercial basis. The situation is worrisome as in the near future, under the next gas package, a market will need to be created for renewable and low carbon gases to decarbonise the gas grid. Continued public support for natural gas, and resistance to continue to provide fossil fuel subsidies could slow down this transition. On the other hand, the EU ETS did play in favour of natural gas, and it could also make renewable gases more competitive.

In the medium term, it appears that a new generation of CEF funding will be necessary. As part of the next long-term EU budget (2021–2027),[98] the EC proposed to renew the CEF to continue to

---

[98]Proposal for a Regulation of the European Parliament and of the Council Establishing the Connecting Europe Facility and repealing Regulations (EU) No. 1316/2013 and (EU) No. 283/2014, COM (2018) 438 final, 06.06.2018.

support the development of, among others, energy infrastructure within trans-European networks. The new budget is estimated to be €42.3 billion and does have an ambition, compared to the previous CEF, namely to accelerate energy sector decarbonisation. The share of the energy sector in the total budget of CEF appears to be €8.7 billion and it should work towards making the European energy system more connected. How much of this amount will be spent on funding LNG (mainly for US LNG), natural gas infrastructure, storage or for much needed research and development for biomethane, hydrogen and CCS remains to be seen.

The CEF 2021–2027 proposals ensures the possibility of using grants for combination with financing from the world's biggest public bank, the EIB, alongside National Promotional Banks or other development and public financial institutions as well as from private-sector finance institutions and private-sector investors, including through Public Private Partnerships. However, on July 2019, the EIB proposed to fully ditch funding fossil fuel projects, when it announced its draft energy lending policy.[99] The draft notes that it will not be possible to present energy projects reliant on fossil fuels (e.g. oil and gas production, infrastructure primarily dedicated to natural gas, power generation or heat based on fossil fuels) to the EIB for approval beyond the end of 2020. And as a result, all the activities of the EIB in the energy sector will be fully aligned with the Paris Agreement. The EIB have been supporting a number of natural gas projects in the past, and recently spent €1.5 billion on the Trans Adriatic Pipeline. As a next step, this proposal was presented to the EU finance ministers in September 2019,[100] however, the Bank has postponed taking a decision on whether to stop financing fossil fuel projects to November as some of its board members would like to continue to financing certain gas projects for a period. If the decision to fully omit all fossil fuel projects is adopted by the EIB's board in November it can have significant implications for the future of gas in the EU energy mix.

---

[99]EIB Energy Lending Policy Supporting the Energy Transformation, 26.07.2019, available at: https://www.eib.org/attachments/draft-energy-lending-policy-26-07-19-en.pdf.

[100]EU Observer news release, EIB fails to end fossil-fuel financing from 2020, 15 October 2019, available at: https://euobserver.com/tickers/146272.

4.11 Chapter Conclusions

In 2015, when the Energy Union was launched, the Climate Action and Energy Commissioner Miguel Cañete stated that it would be partly built through lines of private finance guaranteed by public money.[101] These blended approaches were seen as being central to the EU's flagship Investment Plan. The purpose of using public seed money is to encourage private investment across the five pillars of the planned Energy Union: energy security, renewables, energy efficiency, internal energy market, and research and innovation. The EU financial support mechanisms studied in this chapter have brought long-term and low-cost loans. However, its ultimate goal of establishing markets, i.e. allowing energy to freely flow across borders, has not been as achieved, as currently there is no natural gas pipeline that has been selected as a PCI, and which has not been subsidised.

The EU policies are inconsistent: subsidies for fossil fuels continue alongside tighter regulations against GHG emissions. Ideally natural gas projects should not be subsidised through the PCI, as of now. In previous chapters of this book, it is well acknowledged that there will not be any drastic reduction in the need for natural gas before the 2030s, and the import needs, including infrastructure needs will continue. There are plans in the EU to increase LNG import capacity for instance (as part of diversification policy). It is important to give the right signals for upstream investors and stimulate the market, as natural gas can indeed play a transition-fuel or bridging role for the next two decades, and gradually be replaced by renewables gases. However, there is so much uncertainty over current gas infrastructure, CCS and biomethane that they all bring ambiguity to the table, and they too may require new or adapted infrastructure. Financial support will also be necessary to encourage research and development as well as deployment and commercialisation of renewable gases for use in heating, industrial processes,

---

[101]News Article, Euractive, Energy Union Will Be Built on Public Guarantees for Private Investment, Says Cañete, 17.02.2015, available at: https://www.euractiv.com/section/energy/news/energy-union-will-be-built-on-public-guarantees-for-private-investment-says-canete/.

power generation and transport. As of yet, the EU does not have a silver bullet for these variable objectives. The EU ETS has been delivering GHG reduction, but the question remains whether it will be useful in delivering technologies. It is not yet known, as there is no market for low carbon products downstream for demand, and the marketability of hydrogen is not there. As a starting point, however, redirecting some of the fossil fuel subsidies towards cleaner and renewable gases could instigate a rapid roll out.

The seeds of solutions to problems should be planted ahead of their emergence. Does the EU regulator have a forward-looking risk assessment that responds to those risks facing the completion of the internal gas market and the future role of gas in the EU energy mix? What are the implications of the current regulatory and policy environment of the EU on infrastructure development? How do the rules affect potential suppliers and supply routes? What considerations influence project sponsors decision to invest? What happens if there are no more investments in the gas infrastructure in the period until 2030? Is there enough time to make final investment decisions to decarbonise the gas grid before 2050? The next two chapters will focus on the current state of gas infrastructure in the EU and the decision-making process for investment.

5

Now and Then: The Future of Gas Infrastructure in the EU

5.1 Introduction

This chapter provides an overview of the regulatory framework for development of gas infrastructure in the EU, focusing on cross-border pipelines, interconnections and LNG projects built before and after the Third Energy Package and the Network Codes. It also analyses how the regulatory framework and policy choices of the EU compare with international level standards, for example the UNFCCC Framework and the Paris Agreement. It will explain the state of play, and the impact of proposed amendments to the Gas Directive on energy transition objectives with an assessment of the available policy and regulatory options for development of infrastructure for renewable gases and hydrogen at scale for a timely decarbonisation of the gas grid.

As a majority of gas consumed in the EU is imported and distributed by pipelines, and gas pipelines have been a major recipient of EU public funding over the last decade, this chapter will predominantly present case examples of cross-border pipeline infrastructure. However, LNG and future hydrogen networks will also be covered, as the share of LNG has increased considerably in recent years. Significant private and public

© The Author(s) 2020   **139**
G. Mete, *Energy Transitions and the Future of Gas in the EU*, Energy, Climate and the Environment, https://doi.org/10.1007/978-3-030-32614-2_5

resources have been invested in small and large diameter pipelines, not only in the EU but globally, as they provide supply security and play a major role in balancing the uneven distribution of natural gas resources around the world. Indeed, the global gas pipeline infrastructure market is expected to reach US$1.5 trillion by 2025,[1] as a result of the growth in production and technological advancements in offshore gas fields.

However, gas pipelines are inflexible in terms of volumes as they inevitably tie specific production with specific markets (i.e. compared to oil markets, substitute quantities are not readily and immediately available in natural gas markets). Natural gas infrastructure is an immobile asset, meaning that once made the investment cannot be relocated. Cross-border pipelines, which are fixed and involve multiple jurisdictions, face a significant number of non-commercial risks on top of commercial ones (such as rate of return). Supplies could be interrupted due to technical problems, extreme weather events, terror attacks and other unforeseen factors and circumstances. Pursuing such a project is therefore a high-risk activity, and historically, cross-border natural gas pipelines have experienced detrimental interruptions.[2] This is due to interstate conflicts resulting from the involvement of multiple jurisdictions and diverging economic and political interests. The legal and fiscal stability of the host state plays an important role in investors' decisions to be involved in a certain project.

A robust risk management strategy must be employed when dealing with mega infrastructure projects like pipelines. One way to minimise such risks, as highlighted in the previous chapter, is to use debt financing. In other words, using borrowed money externally raised to fund the sunk, upfront, capital costs. This money is raised in capital markets and then paid back over a defined period of time with project profits. However, as there are high transaction and technical costs involved in the design of natural gas infrastructure financing, the cost of raising capital also tends to be high. In this sense, only projects that

---

[1]Global Market Insights, Gas Pipeline Infrastructure Market Size by Application, January 2019, available at: https://www.gminsights.com/industry-analysis/gas-pipeline-infrastructure-market?utm_source=globenewswire.com&utm_medium=referral&utm_campaign=Paid_globenewswire.

[2]Paul Stevens, *Transit Troubles Pipelines as a Source of Conflict* (A Chatham House Report, 2009).

could prove profitability, a stable rate of return and legal and political predictability will go ahead. For instance, the Nabucco project failed to attract project financing due to a lack of firm commitment for natural gas supplies (partly due to inadequate resources or the will to supply) by prospective producers. Ultimately, the project did not materialise as it could not convince lenders that the sunk costs would be recovered during the lifetime of the project, which therefore carried the risk of becoming a stranded asset.[3]

In a contemporary cross-border pipeline project development, the pipeline itself acts as a SPV that develops, owns and operates the pipeline, ships the gas to market, carries out transport functions and receives a commercial tariff. The SPV acts as the borrower of the debt. Apart from the project's own legal framework (the international and national framework directly applicable to the project), the financial viability of a pipeline project will further be impacted by national issues such as local regulations on environmental matters, labour, health and safety standards and laws on land acquisition and resettlement.

As discussed in Chapter 3 of this book, there will be need for further gas infrastructure in the EU, in particular in the Baltics and South East Europe in the post-2020 period to complete the internal market and further investments in import facilities may be necessary to close the anticipated gas deficit as a result of declining production. The Commission's Sustainable Energy Security Package confirms that natural gas will have to be imported from different sources, including Russia, Norway, Qatar and other producers.[4] The 10 key priorities of the new Commission (taking office in November 2019), continues support for gas (and electricity) and such infrastructure investments are highlighted as a means to deliver a more competitive energy sector and a better deal for European consumers.[5]

---

[3]Andrey Konoplyanik, Energy Charter Treaty: Past, Present and Future, *OGEL*, Volume 3 (2012).

[4]European Commission—Press release, Towards Energy Union: The Commission presents sustainable energy security package, Brussels, 16 February 2016, available at: http://europa.eu/rapid/press-release_IP-16-307_en.htm.

[5]Factsheets on the Commission's 10 priorities, 07.05.2019, available at: https://ec.europa.eu/commission/sites/beta-political/files/factsheets_on_the_commissions_10_priorities_v14.pdf.

These new cross-border pipelines and LNG import facilities will require billions of euros of investment in infrastructure. Chapter 4 explained that the energy sector receives financing from various sources: from investors, through State funding, or via external sources, such as bank loans or funds raised in the capital markets. The selection of a financing method depends on, inter alia, the project timeline, ownership structure, and the regulatory environment, estimated rate of return, the actors involved and the risk factor. As previously discussed, that there are two types of funds available in the EU: finances for commercial projects and funds for strategic projects. If the necessary finances cannot be raised in capital markets and the project cannot be developed on a commercial basis, it may be developed with public funds. As noted by a former regulatory development manager for the Spanish utility Enagas, *"while security of supply may be a very good incentive, no one will pay for it. These are very heavy strategic and political reasons to build infrastructure."*[6]

It remains to be seen whether new sources of finance can be targeted to support the gas infrastructures needs of the EU in the current policy and regulatory environment (with a new gas sector package in the pipeline) and considering the EIB's recent decision to seize financing for natural gas projects as of 2020. Building on the previous chapters on regulation, market and finance principles, this chapter provides a closer look at the implications of policy and regulation on infrastructure design and development, setting out projections on the future of gas infrastructure in the EU.

5.2   Cross-Border Import Pipelines

Focusing on the EU energy mix, Chapter 3 revealed that the union is dependent on natural gas imports. This section provides examples of how the evolving EU regulatory and policy space impacts those gas

---

[6]EU exposed: The shared cost of security of supply, Interview with Gas Strategies, May Issue 2014, available at: http://www.gasstrategies.com/information-services/gas-matters/eu-exposed-shared-cost-security-supply.

projects that connect the EU with third country producers and transit states. As approximately a quarter of these imports come from Russia, the case examples below will start with projects mainly funded and supplied by Russia, before moving to other existing and potential suppliers. Subsequent sections will discuss interconnector and LNG facilities.

5.2.1 North Stream Pipeline

The North Stream pipeline have already been referred to in the previous chapters of this book. This section investigates the legal and financial structure of the pipeline in more detail *vis a vis* EU energy law and policy in-order to identify where areas of interests have conflicted, as well as how this has impacted on natural gas supplies to European consumers and the market. The North Stream pipeline has two sections. The first section of the pipeline, a twin pipeline system referred as North Stream 1, has been in operation since 2012 with a combined total capacity of 55 bcm of gas a year (sufficient for circa 26 million European households).

The North Stream 1 pipeline, which crosses the Baltic Sea, was one of very few projects not based on an international treaty between the states most interested in it, in this case Russia and Germany. Instead, the legal foundation of North Stream is formed by a private law contractual framework.[7] The pipeline owner and operator, the international consortium Nord Stream AG, has applied, and will continue to comply with relevant international legal principles, including provisions of the United Nations Convention on the Law of the Sea (UNCLOS),[8] some regional multilateral environment agreements and EU law, as well as applicable national legislation.[9]

---

[7]Sergei Vinogradov, Challenges of Nord Stream: Streamlining International Legal Frameworks and Regimes for Submarine Pipelines, *German Yearbook of International Law*, Volume 52 (2009), p. 241.

[8]UN General Assembly, Convention on the Law of the Sea, 10 December 1982.

[9]Sergei Vinogradov and Gokce Mete, Cross-Border Oil and Gas Pipelines in International Law, *German Yearbook of International Law* (FOCUS: International Energy Law), Volume 56 (2013).

The financing agreements for the second phase of the pipeline project—North Stream 2—was signed in April 2017 between the project company and ENGIE, OMV, Royal Dutch Shell, Uniper, and Wintershall. These five European energy companies will provide long-term financing for 50% of the total cost of the project.[10]

The North Stream pipeline (1 and 2) is a project specifically contemplated and designed to bypass other states' territories. A decision was taken by the supplier to avoid transit through Ukraine and provide natural gas directly to Northern Europe. Gazprom's transit contracts with Ukraine are due to expire after December 2019 and, while Gazprom has decided not to renew them in light of the various disputes on transit fees and gas prices, the CEO of the company announced in June 2019 that Gazprom is *"ready to extend the transit contracts in economically profitable conditions in case of the righteous solution of all litigations between Gazprom and Naftogaz"*.[11] In the past, these disputes have repeatedly resulted in the reduction of transit flows via Ukraine. However, the EU does not view the North Stream 2 pipeline as contributing to the energy security of the EU as it is not regarded as a diversification of sources but only of routes (the source remains the same, Russian gas supplied by Gazprom), which means volumes contracted by European gas buyers will be diverted to the North Stream 2 instead of the transit route via Ukraine, Poland and Slovakia.

Chapter 2 of this book discussed the problems faced by the North Stream 1 on-land section (OPAL). The OPAL section of the pipeline in Germany did receive an exemption from the then applicable Second Energy Package. Subsequently, this exemption was capped by the Commission, with only 50% rights to the pipeline capacity. Gazprom could not use this capacity despite there being no other interested shippers. Hence, half of the pipeline capacity ran empty. In October 2016, the Commission decided to allow OPAL to utilise its full capacity, a

---

[10]ENGIE, Press release, 24.04.2017, Nord Stream 2 AG and European Energy Companies Sign Financing Agreements, available at: https://www.engie.com/wp-content/uploads/2017/04/press-release-nord-stream-2-project-20170424.pdf.

[11]News Article, Energy Reporters, Ukraine Looks to Renew Gazprom Transit Deal, 25.07.2019, available at: https://www.energy-reporters.com/consumption/ukraine-looks-to-renew-gazprom-transit-deal/.

move which was contested by the Polish Company PGNiG. However, in July 2017, the German Higher Regional Court and the European Court of Justice dismissed these claims and upheld the Commission's decision to lift the capacity cap on Gazprom, allowing the company to increase use of the OPAL pipeline capacity from to 80 to 100%.[12]

The Commission intervened with the North Stream 2 project. The Legal Service of the Commission alleged that the pipeline's offshore section crossed a number of EU Member States (Finland, Sweden, Denmark and Germany), and as EU law and Russian law on energy differ, there was the potential for a conflict of laws, which *"would best be resolved through international negotiations"*.[13] The Commission have also been seeking to renegotiate the North Stream 2 project on behalf of Germany. However, as there is no intergovernmental agreement in place between Russia and Germany, there is no legal ground to allow the Commission to take any action in this direction. The European Parliament spent several sessions looking for a solution to bring the project in line with EU law, applying non-discriminatory tariffs and making capacity available to third parties.

The concern raised was that the offshore pipeline would be constructed not on the basis of the EU law but Russian law, as the Third Energy Package does not extend to offshore pipelines. The Commission's DG-Energy was of the opinion that EU law should be extended to pipelines outside the EU territory. It was proposed that this be achieved via an amendment to the Gas Directive, to make the Gas Directive applicable to pipelines to and from third countries, including existing and future pipelines, up to the border of EU jurisdiction. This move to change the rules applicable to existing and future pipelines brought legal uncertainty for natural gas pipeline projects, particularly

---

[12]News release, Tass, German Court Lifts Provisional Measures Limiting Gazprom Access to OPAL pipeline, available at: http://tass.com/economy/958228.

[13]Brussels Admits EU Law Does Not Apply to Nord Stream 2, Special Reprint from International Oil Daily for Oxford Institute of Energy Studies, 21.09.2017, available at: https://www.oxfordenergy.org/wpcms/wp-content/uploads/2017/06/Brussels-Admits-EU-Law-Does-Not-Apply-to-Nord-Stream-2.pdf.

with respect to their bankability. A leaked letter, dated 12 September 2017, revealed that the Legal Service of the Commission concluded that the application of Third Energy Package regulations to offshore import pipelines, such as Nord Stream II, *"would raise specific legal and practical questions, arising... from the fact that Union rules cannot be made unilaterally binding on the national authorities of third countries"*.[14] Nevertheless, in the context of his State of the European Union Speech of the 13 of September 2017, then President Juncker announced that *"building upon the solidarity aspect of the Energy Union, the Commission will propose common rules for gas pipelines entering the European internal gas market"*.[15] The leaked letter was addressed to Jerzy Buzek, Chairman of the European Parliament's Committee on Industry, Research and Energy, and signed by the Commission's Vice President Maroš Šefčovič and EU Energy Commissioner Miguel Arias Cañete.

In addition to this, a leaked legal opinion of the EU Council's Legal Service, released on the 1 March 2018, stated that the Commission's proposal to make all import pipelines crossing Member States' Exclusive Economic Zones (EEZ) comply with the EU law *"lacks any reasoning on the regulatory power of the Union over offshore pipelines"* as these are regulated under UNCLOS. Hence, the EU does not have jurisdiction to apply energy law to pipelines crossing the EEZ of Member States.[16]

The proposal passed in 2019 with a compromise, following extensive discussions.[17] The trilogue between the Parliament, the Council and

---

[14]Brussels Admits EU Law Does Not Apply to Nord Stream 2, Special Reprint from International Oil Daily for Oxford Institute of Energy Studies, 21.09.2017, available at: https://www.oxfordenergy.org/wpcms/wp-content/uploads/2017/06/Brussels-Admits-EU-Law-Does-Not-Apply-to-Nord-Stream-2.pdf.

[15]Full speech available at: https://ec.europa.eu/avservices/video/player.cfm?ref=I143451.

[16]News Article, Reuters, EU Legal Blow to Bid to Regulate Russia's Nord Stream 2 Pipeline, 05.03.2018, available at: https://uk.reuters.com/article/uk-eu-gazprom-nordstream/eu-legal-blow-to-bid-to-regulate-russias-nord-stream-2-pipeline-idUKKBN1GH28L.

[17]Final text of the Gas Directive Amendment is available here: https://data.consilium.europa.eu/doc/document/PE-58-2019-INIT/en/pdf.

the Commission, agreed that the application of Gas Market Directive would be restricted to the territory and territorial sea of the Member State where the first interconnection point is located. This avoids conflict with UNCLOS; however there are still uncertainties and problems with this amendment with potentially serious implications on existing, planned and future gas pipeline infrastructure. These will be discussed further in Chapter 6 in relation to the decision-making framework. In order to lay the necessary groundwork, this chapter will provide an understanding of the intersection of energy and climate regulation and energy supply and infrastructure investment security.

When it comes to EU energy security, the future of North Stream 2 plays an important role indeed. Although the volume of gas traversing through Ukraine to the EU has been decreasing as a result of North Stream 1, this route still carried 84.3 bcm of Russian gas in 2017, accounting for 24% of the EU's total natural gas imports.[18] Any reduction or even complete halt of the transit of Russian gas supplies from Ukraine would have significant impact on security of supply and, consequently, would mean Ukraine losing nearly US$2 billion in annual transit fees, something that the Commission fiercely opposes. In the world of project finance and open markets, however, the producer should be allowed to design its supply projects in a risk-free manner that increases the chances of the project's bankability. In relation to the transit route via Ukraine, Gazprom has experienced high transit risks, litigation and disruptions in the past. The economics of renewing transit contracts via Ukraine, in the face of Ukraine's aging natural gas transportation system, meant that constructing a second phase to the North Stream 1 pipeline already in operation made good business sense. However, with the recent Gas Directive Amendment in 2019, the future of North Stream 2 remains uncertain. This is compounded by the pending international investment arbitration case brought by North Stream AG against the Commission under the Energy Charter Treaty (ECT).[19]

---

[18]Jack Sharples, Ukrainian Gas Transit: Still Vital for Russian Gas Supplies to Europe as Other Routes Reach Full Capacity, OIES, May 2018.

[19]Politico, New Article, Nord Stream 2 sues the EU over new gas rules, 26.09.2019 available at: https://www.politico.eu/article/nord-stream-2-sues-the-eu-over-new-gas-rules/.

5.2.2 South Stream Pipeline

The South Stream pipeline project is another example of the dynamics of decision-making practice in gas infrastructure projects in the EU. This cross-border pipeline has now been scrapped but the events, which led to its cancellation, offer some valuable inputs to understand how international law and EU energy law have interacted with the implementation of this project. The South Stream pipeline was designed as a series of connected national pipelines crossing multiple EU Member State jurisdictions. The pipeline consisted of an offshore section across the Black Sea and terrestrial pipelines across several countries in Southeastern Europe. The offshore section was to run through the EEZ of Russia, Turkey and Bulgaria and was meant to be developed by an international consortium, South Stream Transport B.V. Two separate agreements to build the submarine section were signed by Russia with Turkey and Bulgaria. In 2013, a draft Environmental Impact Assessment (EIA) report was submitted in accordance with permitting procedures under Turkish law[20] and the Bulgarian EIA approving the investment proposal for the pipeline project was published in February 2014.[21]

The onshore section was meant to cross Bulgaria, Hungary, Slovenia and Italy, with gas branches going to Croatia and Serbia. A set of bilateral IGAs between Russia and the respective countries were signed between 2008 and 2010. The IGAs were signed between Russia and six EU Member States separately[22] as well as with Serbia, an Energy

---

[20]Press release, Gazprom, South Stream Offshore Pipeline in Turkish Waters Will Have no Negative Environmental Impact, 25.12.2013, available at: http://www.gazpromexport.com/en/presscenter/news/1185/.

[21]South Stream Offshore Pipeline—Bulgarian Sector Addendum to the Environmental and Social Impact Assessment (ESIA), October 2014, available at: https://www.south-stream-transport.com/media/documents/pdf/en/2014/10/ssttbv_addendum-to-the-environmental-and-social-impact-assessment-esia-south-stream-offshore-pipeline-bulgarian-sector_335_en_20141030.pdf.

[22]Individual intergovernmental agreements signed with Bulgaria, Hungary, Greece, Slovenia, Croatia and Austria.

Community Member, which is also bound by the EU energy *acquis*.[23] These were supplemented by agreements on cooperation between Gazprom and authorised national companies. Several joint companies based on essentially equal shares were established by Gazprom with local partners for every on-land section. These companies were registered and incorporated under the applicable legislation of each participating State. Gazprom and Austrian OMV intended to set up a similar joint project company, for the design, financing, construction and operation of a spur to Austria. Some typical legal issues such as tariffs and transit fees, land use rights and access regime were to be governed by the national regulations of the State traversed by the pipeline.

The Commission challenged the provisions of the bilateral intergovernmental agreements signed to implement the project. Accordingly, the Commission investigated the terms of the international pipeline agreements and found that they did not comply with the Third Energy Package. The Commission particularly opposed the ownership structure of the proposed TSO Gazprom, which was to own both the pipeline infrastructure and the gas to be exported. The Third Gas Directive requires any TSO (including third-country companies) operating in the EU to be certified and structured under one of the three types of unbundling models foreseen in the legislation.[24] In essence, unbundling obligations are deemed necessary to ensure that the commodity owner is unable to distort or restrict free access to the transmission system, thereby creating market barriers within the internal market. In addition, Gazprom's right to set the transmission tariffs (instead of the regulator as foreseen in the Third Energy Package), and the fact that it is the only shipper, raised further tensions between the IGAs terms and the regional legislation.

It is important to point out that since 2012, the Commission has had the power to participate in the negotiation of the Member States' energy related IGAs with third countries as an observer (upon request

---

[23]The Energy Community Membership entails expansion of the core EU energy acquis as adapted to the special circumstances in the Member State in question.

[24]The ownership unbundling, the independent system operator or the independent transmission operator, Articles 9 and 11 Gas Directive.

of the Member State or of the Commission itself subject to the Member State's written approval). The Commission also has the power to provide advice on how to circumvent any inconsistency between EU law and IGAs.[25] This was exactly what was sought by the Commission in the South Stream case, initially calling for renegotiation of the existing agreements. This was done to bring them in line with the EU energy *acquis* and was subsequently accorded the mandate to renegotiate the bilateral contracts with Russia on behalf of the Member States concerned.

Against this, Russia argued that the IGAs were valid and enforceable under international law and that they take precedence over EU law. The EU Member States along the pipeline corridor were therefore caught up in a legal conundrum. If they were to withdraw from or breach their respective agreements with Russia, could this result in an interstate dispute? Whereas continuance of the project, according to the IGAs, could result in the Commission initiating infringement procedures against them for the alleged violation of EU law. The paradox became irrelevant, as Russia cancelled the pipeline project and proposed an alternative route through Turkey to be used for domestic supplies and transit to southern Europe. This so-called 'Turk Stream' pipeline project is a way for Russia to sell its gas to Europe without being subject to EU rules, as Turkey is not an EU member. In fact, Turk Stream is not about connecting Turkey with Russia anymore, it is now often referred to as South Stream reloaded.

In 2015, the EU promoted greater involvement of the Commission in bilateral energy agreements in the Energy Union Communication document, released on the 25 February 2015,[26] as a means to minimise and eliminate market distortions. In this document, the Commission

---

[25]Decision No. 994/2012/EU of the European Parliament and of the Council of 25 October 2012 establishing an information exchange mechanism with regard to intergovernmental agreements between Member States and third countries in the field of energy [2012] OJ L299/13.

[26]Energy Union Package Communication from the Commission to the European Parliament, the Council, the European Economic and Social Committee, the Committee of the Regions and the European Investment Bank; A Framework Strategy for a Resilient Energy Union with a Forward-Looking Climate Change Policy, COM (2015) 80 final, p. 7.

made public its intention to take a further step and obtain the right to 'actively participate' in natural gas contract negotiations concluded between Member States and third-country suppliers. No contacts have so far been negotiated by the Commission with third countries, and it remains open how any such mechanism will function regarding sensitive issues, such as national sovereignty and commercial confidentiality. The evolution criteria and further details, such as whether this mechanism will cover not only IGAs but also private gas supply contracts, still awaits elaboration. It must be noted, however, that although Article 194 of the TFEU confers on the EU the power to take legislative action in the energy field, the second paragraph of TFEU allows opt outs. Member States can decide on their energy mix and may block legislative action by the Commission.

This and the North Stream 2 pipeline case demonstrates that the interaction between international law and regional law in the energy sector continues to cause tension within the EU. This issue remains a continuous problem, which is handled differently in different areas of the energy market legal systems, and results in legal uncertainties.

5.3   The World Trade Organisation (WTO) Dispute Between Russia and the EU

The legal dispute around EU energy law versus international law has also been brought up in a WTO dispute raised by Russia in 2014.[27] Russia requested consultation with the EU and Member States regarding measures relating to the energy sector under the Third Energy Package Directives, regulations, implementing legislation and decisions. The request was based on the claim that the application of the Third Energy Package for non-EU countries breaches the WTO's General Agreement on Trade in Services (GATS), the General Agreement on Tariffs and Trade (GATT) 1994, as well as the Agreement on Subsidies

---

[27]WT/DS476/2.a, European Union and Its Member States—Certain Measures Relating to the Energy Sector, available at: https://www.wto.org/english/tratop_e/dispu_e/cases_e/ds476_e.htm.

and Countervailing Measures and the Agreement on Trade, the Agreement on Trade Related Investment Measures, and the Agreement Establishing the WTO.

The core of the claim was that several aspects of the Third Energy Package, including a requirement on granting access to natural gas and electricity networks and unbundling requirements for third country suppliers, unjustifiably restricted imports of natural gas originating in Russia and discriminated against Russian natural gas pipeline transport services and service suppliers. The EU's response to the claim was that as the rules apply not only to Russia but to all non-European and European companies, the measures under the regulation are not discriminatory.[28]

Russia further claimed, with respect to the South Stream pipeline, that Article 34 of the 1997 EU-Russia Partnership and Cooperation Agreement, should exempt it from the Third Energy Package as under this agreement *"the parties shall use their best endeavours to avoid taking any measures or actions, which render the conditions for the establishment and operation of each other's companies more restrictive than the situation existing on the day preceding the date of signature of the Agreement"*.[29] Russia also asserted that the Third Energy Package was backdated to Gazprom's supply contracts signed prior to its entry into force in 2009.

The consultation failed to resolve the dispute, and Russia requested that the Dispute Settlement Body establish a Panel to examine the matter, a request subsequently rejected by the EU. Russia then made a second request which could not be denied. Therefore, in July 2015, a Panel was established. It was composed by the Director-General of the Dispute Settlement Body and a final report was expected before the end of 2017. However, a revised timetable was adopted after consultation with the parties due to the complexity of the issue.

---

[28]WT/DS476/2.a, European Union and Its Member States—Certain Measures Relating to the Energy Sector, available at: https://www.wto.org/english/tratop_e/dispu_e/cases_e/ds476_e.htm.

[29]Partnership and Cooperation Agreement establishing a partnership between the European Communities and their Member States, of the one part, and the Republic of Uzbekistan, of the other part—Protocol on mutual assistance between authorities in customs matters, 31.08.1999, OJ L 229.

On 10 August 2018, the WTO panel report was circulated to Members.[30] Russia lost the bulk of its claim against the EU gas pipeline rules under the ruling including those concerning the rules on unbundling.[31] The WTO panel did not find any basis to Russia's claim concerning alleged EU discrimination in its Third Energy Package against Russian pipeline transport services, service suppliers, or against Russian natural gas. However, the panel raised a number of issues on the WTO-compatibility of EU energy policy. In particular, the panel upheld Russia's complaint about an unbundling exemption for Germany's OPAL pipeline and also agreed that Croatia, Hungary and Lithuania had discriminated against Russia by requiring a security of energy supply assessment for foreign, but not domestic, pipeline operators.

From an international law perspective, the outcome is important as the WTO panel rejected the EU's objection that it would not be appropriate for the WTO panel to address differences among the national laws of various EU member States where these are accorded discretion in implementing an EU legal instrument.[32]

5.4   TANAP and TAP

On the 25 October 2011, Azerbaijan and Turkey signed an IGA in relation to the *"Sale of Natural Gas to Turkey and the Transit Passage of Natural Gas through Turkey and the Development of a Standalone Pipeline for the Transportation of Natural Gas across the Territory of Turkey"*.[33] The two States then signed another IGA, on the 26 June 2012, specifically

---

[30]WT/DS476/2.a, European Union and Its Member States—Certain Measures Relating to the Energy Sector, available at: https://www.wto.org/english/tratop_e/dispu_e/cases_e/ds476_e.htm.

[31]European Commission—Press release, Commission welcomes WTO ruling confirming lawfulness of core principles of the EU third energy package, 10.08.2018, available at: http://europa.eu/rapid/press-release_IP-18-4942_en.htm?locale=en.

[32]For a detailed discussion on the role of WTO rules in international regulation of natural gas transport, see Vitaliy Pogoretskyy, Freedom of Transit and Access to Gas Pipeline Networks under WTO Law (Cambridge University Press 2017).

[33]Joint Opinion of the Energy Regulators on TAP AG's Exemption Application, 17, available at: https://www.autorita.energia.it/allegati/docs/13/249-13all.pdf.

for the TANAP Pipeline, which will transport natural gas from the Eastern border of Turkey to the Western border, stretching almost 2000 km.[34]

The Trans-Adriatic Pipeline (TAP)[35] will then connect to TANAP at Kipoi, in Greece, on the border with Turkey. From there, the TAP will cross Greece and Albania towards the Adriatic coast. The first gas deliveries from TANAP to Turkey are expected in 2019 and, approximately a year later, to Europe.

The TANAP project implements the Southern Corridor initiative, a US-backed plan to enable Europe to diversify its delivery routes away from Russia. This initiative dates back to the construction of the Baku-Tbilisi-Ceyhan[36] oil pipeline in the early 2000s that brought Azeri oil to international markets without the need for transit through Russia. The TANAP project, however, is becoming increasingly expensive, with total costs currently estimated at around US$45 billion. And in contrast to initial expectations, the State Oil Fund of the Republic of Azerbaijan (SOFAZ) and BOTAŞ, the Transmission System Operator of Turkey, will not be able to jointly fund the project from their budget, as SOFAZ's funds are running low due to the 2014 oil price slump.

Therefore, BOTAŞ received a loan of US$400 million from the World Bank,[37] a US$500 million loan from the EBRD,[38] a US$600

---

[34]Agreement Between the Government of the Republic of Turkey and the Government of the Republic of Azerbaijan Concerning the Transit Passage of Natural Gas Originating and Transiting from the Republic of Azerbaijan to and through the Territory of the Republic of Turkey via the Trans Anatolian Natural Gas Pipeline System, 26.06.2012, available at: http://www.tanap.com/content/file/TANAPIGA.pdf.

[35]Factual information on TAP can be found at the project's home page available at: https://www.tap-ag.com/.

[36]The BTC pipeline transports oil from the Azeri-Chirag-Gunashli field from Shah Deniz across Azerbaijan, Georgia and Turkey. It links the Sangachal terminal on the shores of the Caspian Sea, to the Ceyhan marine terminal on the Turkish Mediterranean coast. In total length, the BTC pipeline is 1768 km long.

[37]World Bank, Press release, Trans-Anatolian Natural Gas Pipeline Project (TANAP) 20.12.2016, available at: http://www.worldbank.org/en/news/loans-credits/2016/12/20/trans-anatolian-natural-gas-pipeline-project-tanap.

[38]News Article, Reuters, EBRD Board Approves $500 Million Loan for TANAP Gas Pipeline Project, 18.10.2017, available at: https://uk.reuters.com/article/europe-gas-ebrd/ebrd-board-approves-500-mln-loan-for-tanap-gas-pipeline-project-idUKL8N1MR4YS.

million loan from the Asian Infrastructure Investment Bank (AIIB)[39] and a €932 million loan from the EIB.[40] Among various proposals (Nabucco, White Stream, etc.) the final investment decision was taken in 2013 on the TAP, connecting TANAP through Turkish territory. As mentioned in Chapter 2, an exemption from the Third Energy Package rules on tariff, third party access and certification of the TSO was granted for 25 years for the TAP project. Otherwise, the project would not be commercially viable or attractive for private investors (pipeline consortium) and external financing.

The project HGA grants exclusive rights to the project entity in determining allocation of capacity, the calculation method and the amount of transit fees.[41] TANAP does not cross borders within the EU to deliver gas, so contractual mismatches are not a primary concern here. Since Turkey is not an EU Member State, nor a member of the Energy Community, Azerbaijan with its 51% stake in the project will enjoy full control of the transport of gas flows via TANAP. Transport tariffs and third-party access is also to be negotiated between Turkey and Azerbaijan, a privilege that Russia does not have over transit pipelines in Ukraine for instance which carries Russian gas.

The buyers of this gas include: Bulgargaz (around 1 bcm), DEPA (around 1 bcm), E.ON (around 1.6 bcm), and Enel, Gas Natural, GDF SUEZ, Shell, Axpo Trading, and Hera Trading.[42] Once TANAP is connected to TAP, Italy will gain a new terminal, and both Greece and Albania will have interconnectors; all three would have the capability of becoming transit points as the volume of gas shipments increases.

---

[39]Asian Infrastructure Investment Bank, Press release, AIIB Approves $600 Million to Support Energy Project of Azerbaijan, 20.12.2016, available at: https://www.aiib.org/en/news-events/news/2016/20161222_001.html.

[40]News Article, *Hurriyet Daily*, European Investment Bank Approves 932 Million Euro Loan for TANAP Pipeline, 16.03.2018, available at: http://www.hurriyetdailynews.com/european-investment-bank-approves-932-mln-euro-loan-for-tanap-pipeline-128844.

[41]Art. 3 TANAP HGA.

[42]News release, Enerdata, Shah Deniz Consortium Signs 10 bcm/year Gas Supply Agreements, 23.09.2013, available at: https://www.enerdata.net/publications/daily-energy-news/shah-den-iz-consortium-signs-10-bcmyear-gas-supply-agreements.html.

The initial Southern Corridor idea was to accommodate large volumes of natural resources from the Caucasus, Central Asia and from the Middle East and North Africa. However, over time the US has lost interest due to various factors, but predominantly a lack of availability and accessibility of oil and gas for export to Europe to fill the Corridor. The TANAP project is currently contracted to carry only 16 bcm of gas annually, 10 bcm for Europe and 6 bcm for Turkey. The pipeline's capacity is expected to exceed 31 bcm in 2026. The present share of the Southern Corridor in EU supply security is relatively small, yet it is a concept initially planned to accommodate additional supplies through the TANAP Pipeline from Turkmenistan (via a potential Trans-Caspian Pipeline Project). However, as mentioned previously in Chapter 3, the likelihood of new sources joining the TANAP pipeline from producers in the region, such as Iraq, Iran and Israel are currently very slim.

The diversification policy is not a new item. It has been pursued since the first (2006) and then second (2009) transit disruptions between Russia and Ukraine. In the aftermath of these events, the EU investigated alternative supply sources. This led to the development of the Southern Corridor, a route by-passing Russia to be filled by natural gas sourced from Central Asian, Caspian and Middle East regions.

The Ukrainian crisis of 2014 has no doubt escalated fears that the flow of 'transit' gas can be interrupted. This was a push for the EU to rethink its energy policies towards measures to decrease its vulnerability to disruption and dependency on suppliers. To this end, diversification is not a new policy but one of increasing importance. In fact, the EU initiated this policy in 2008 (as a response to the Russo-Ukraine energy crisis) under the 'EU Energy Security and Solidarity Action Plan: Second Strategic Energy Review'[43]:

---

[43]Communication from the Commission to the European Parliament, the Council, the European Economic and Social Committee and the Committee of the Regions—Second Strategic Energy Review: an EU energy security and solidarity action plan, COM (2008) 781 final, Brussels, 13.11.2008.

A southern gas corridor must be developed for the supply of gas from Caspian and Middle Eastern sources, which could potentially supply a significant part of the EU's future needs. This is one of the EU's highest energy security priorities. The Commission and Member States need to work with the countries concerned, notably with partners such as Azerbaijan and Turkmenistan, Iraq and Mashreq countries, amongst others, with the joint objective of rapidly securing firm commitments for the supply of gas and the construction of the pipelines necessary for all stages of its development. In the longer term, when political conditions permit, supplies from other countries in the region, such as Uzbekistan and Iran, should represent a further significant supply source for the EU.

The feasibility of a block purchasing mechanism for Caspian gas ("Caspian Development Corporation") will be explored, in full respect of competition and other EU rules. Transit for the gas pipelines will need to be agreed with transit countries and notably Turkey in a way that respects both the basic principles of the EU acquis and their legitimate concern for their own energy security. The Commission will invite representatives of the countries concerned to a Ministerial level meeting to secure concrete progress and a timetable to reach agreement. It will seek to identify by mid-2009 any remaining obstacles to the completion of the project which will be the subject of a Communication on the Southern Gas Corridor to the Council and Parliament.[44]

Political will alone is not sufficient to construct a pipeline of this magnitude. An appropriate price, volume guarantees and technical competence by pipeline consortiums were the key elements that determined the final investment decision of the TANAP pipeline, which is not subject to the Third Energy Package rules. As the project has proved to be commercial, it has secured loans from four reliable international finance institutions including the EU's EBRD and EIB.

First commercial gas deliveries to Turkey via TANAP commenced on 30 June 2018 with gas deliveries expected to reach the EU in 2020.[45]

---

[44]Communication from the Commission to the European Parliament, the Council, the European Economic and Social Committee and the Committee of the Regions—Second Strategic Energy Review: an EU energy security and solidarity action plan, COM (2008) 781 final, Brussels, 13.11.2008, p. 4.

[45]News release, Sputnik, First Gas Deliveries to Turkey Launched via TANAP Pipeline, 30.06.2018, https://sputniknews.com/world/201806301065914412-tanap-gas-delivery-turkey/.

The long-sought Southern Corridor will contribute to European energy security and will place Turkey in the transit country position, albeit with little influence to begin with, due to the miniscule volumes of gas to be transported to the EU.

The question arises whether a volume increase could be expected from Azerbaijan considering that natural gas supplies from Azerbaijan to Europe have always been favoured by the West as an alternative to Russian gas, and the fact that the country has been co-operating with Western companies for two decades on upstream projects. For example, the Azeri-Chirag-Deepwater Guneshli oil project and the Shah Deniz gas condensate project, which has been producing natural gas since 2006, are both led by British Petroleum (BP).

Overall, Azerbaijan is a net energy exporter; the country produced around 30 bcm of gas in 2016.[46] There are some gas swaps with Iran as the Nakhchivan exclave is not connected to the pipeline system of Azerbaijan. Natural gas accounted for more than 60% of Azerbaijan's total domestic energy consumption in 2013.[47]

At the end of 2015, Azerbaijan's national oil company SOCAR had already invested about US$10 billion in international projects, and up to US$20 billion will be invested over the next ten years on the Southern Corridor. In addition to the already operating Shah Deniz and Absheron fields, the next projects in Azerbaijan are the Nahcivan and Umid fields, which will increase the export potential of Azerbaijan. According to SOCAR, the Umid field in the South Caspian Sea holds around 200 bcm of gas and 30 million tonnes of gas condensate. Absheron, also located in the Caspian Sea, is estimated to contain 350 bcm of natural gas and 45 million tonnes of condensate. The Umid field is not yet developed as Azerbaijan has not been able to attract foreign investment due to the current price environment since the collapse

---

[46]News Article, *AzerNews*, Azerbaijan Gets Over 30 bcm of Associated Gas from ACG Block, 02.06.2017, available at: https://www.azernews.az/oil_and_gas/114131.html.

[47]US Energy Information Administration, Country Analysis Brief: Azerbaijan, 22.06.2016, available at: https://www.eia.gov/beta/international/analysis.cfm?iso=AZE.

of the oil price in 2015. As far as Absheron is concerned, Total and SOCAR signed an agreement to start production after 2021.[48]

After 2025, more natural gas might become available to fill the TANAP pipeline based on further production from the Shah Deniz field and the development of new fields, such as Umid and Absheron.[49] Hypothetically, natural gas from Turkmenistan could also join the project. For the TANAP project, SOCAR needs to invest US$3.5 billion every year in addition to the US$3–4 billion already invested in the project. Azerbaijani gas could in principle make a significant difference to the small markets in south-eastern Europe, particularly in cases where diversifying away from Russian supply is a policy priority despite the fact the TAP has a small capacity.[50]

For the Southern Corridor to make a significant difference in the context of EU diversification policy, sources of natural gas from Turkmenistan should join the corridor. The question then is: where will the money come from? Over the last fifteen years, Azerbaijan has been cash-positive and sometimes even evidences a double-digit GDP growth. In 2014, GDP growth was 6% despite the global economic situation. According to the IMF and the World Bank, over the next five years Azerbaijan will experience GDP growth on a year by year basis.[51] Taking the oil price into consideration, the Azeri economy will grow on the back of a US$60 per barrel price, and even if that is not the case, the Azeri market is perceived as secure due to the existence of the Sovereign Wealth Fund (SOFAZ) with a balance sheet of around

[48]News Article, Reuters, Fact Box: Azerbaijan's Main Gas Fields, 24.02.2017, available at: https://www.reuters.com/article/us-azerbaijan-gas-factbox/factbox-azerbaijans-main-gas-fields-idUSKBN1630SG.

[49]Simone Tagliapietra and Georg Zachmann, Designing a New EU–Turkey Strategic Gas Partnership, Bruegel, 01.07.2015, available at: http://www.bruegel.org/nc/blog/detail/article/1666-designing-a-new-eu-turkey-strategic-gas-partnership/.

[50]News Article, Natural Gas Europe, Europe's alternatives to Russian, 11.04.2015, available at: http://www.naturalgaseurope.com/analysis-europe-alternatives-russian-gas-23151.

[51]World Bank, Global Economic Prospects: Broad-Based Upturn, but for How Long? January 2018, available at: http://www.worldbank.org/en/publication/global-economic-prospects.

US$45 billion. However, in some years SOFAZ has closed the state budget deficit (€2 billion) but this deficit has continued to grow.

Any investment should undertake a proper feasibility study, not only economically and commercially, but socially as well. In the Southern Corridor, there are many different elements, and detailed feasibility studies have been carried out in the past and Azerbaijan is a major player in the current circumstances as a producer. It has been a reliable partner with a positive track record on investment in all its international partners in international projects and represented positive cash flows. However, looking forward, Azerbaijan's scope of interest with regard to cooperation with the EU is somewhat minimal. For instance, Azerbaijan is refraining from joining The Deep and Comprehensive Free Trade Areas with the EU as this liberalistic market structure does not act in accordance with Azerbaijan's nationalistic economic structure.[52]

5.5   Future of Potential Cross-Border Pipeline Projects

As Chapter 3 on the EU energy mix revealed, there is a rapid decline in domestic gas production in the UK and the Netherlands, which supply natural gas to Northern Europe. European natural gas production is expected to be around 220 bcm by 2023, losing over 40 bcm compared to 2017.[53] Further case studies below will assess non-Russian gas suppliers' perspectives and how EU energy regulation is likely to impact their project viability, and whether new infrastructure investment will be required to address this gap.

One important point that requires explanation from the outset is that resource nationalism is common practice amongst the world's major

---

[52]Farid Guliyev, "After Us, the Deluge: Oil Windfalls, State Elites and the Elusive Quest for Economic Diversification in Azerbaijan", Caucasus Analytical Digest No. 69, 2015.

[53]IEA, Gas 2018, 26.06.2018, available at: https://www.oecd-ilibrary.org/energy/gas-2018_gas_mar-2018-en.

energy producers. Producers exercise close control of their export routes, and their investment policies predominantly seek some sort of ownership and assets in the midstream and downstream sectors that they supply. Producer states naturally have an interest in resource maximisation; they will seek the highest value they can achieve for their resources. From a producer's point of view, longer-term commitments in the supply markets are crucial, particularly in the gas sector, to be able to take an investment decision to develop upstream fields. In Europe, however, the markets are increasingly becoming short-term and thanks to the formation of virtual and physical gas hubs, spot trade and auctions have become widespread.

Among the producer states, which may potentially export further gas supplies to the EU, the Turk Stream pipeline, the long-debated Trans-Caspian Pipeline and the East-Med pipeline have the most likelihood to ever being commissioned, although overall feasibility is dubious due to a variety of factors related to market conditions (including competition with LNG and renewables) and climate policy and regulatory factors (politicisation of the gas sector). The possibility of any gas exports from Iran and Iraq via Turkey to the EU is even less likely to materialise. These will now be discussed below.

5.5.1 Turk Stream

In October 2016, Turkey and Russia signed an IGA for the construction and development of the Turk Stream natural gas pipeline project. In view of the lack of significant demand increases that would require new pipeline infrastructure investment, the pipeline raised questions and has been opposed by the public as it is perceived to increase Turkey's dependence on Russia.

The first line was going to be constructed by an Italian company, Saipem, under the same contract that was concluded for the construction of the scrapped South Stream project. However, as they failed to reach an agreement for a deal worth €2.4 billion, the Swiss-based Allseas has now been awarded a contract to lay the first line of the Turk Stream offshore gas pipeline with its single-lift installation and decommissioning vessel, Pioneering Spirit (the world's largest construction vessel, equipped with six welding stations and six coating stations), starting from the first half of 2017. Construction of the first line of Turk Stream should be completed by December 2019.

Turkey already has a direct natural gas pipeline from Russia, Blue Stream, and a transit pipeline, the Western Line, through Ukraine, Moldova, Romania and Bulgaria. The Turk Stream pipeline project will substitute 14 bcm of Russian gas that currently flows through the Western Line. The Blue Stream direct pipeline, based on a 1997 intergovernmental agreement between Russia and Turkey and commissioned in 2003,[54] has an annual capacity of 16 bcm. The Turk Stream project required an estimated €11.4 billion in investment and, as it is an offshore project, these costs could increase in the implementation stage. Furthermore, there is no codified international law governing the offshore section of the pipeline other than the IGA, as Turkey is not a party to UNCLOS[55] and Russia is not a party to the ECT.[56]

After the cancellation of the South Stream project in 2014, due to regulatory conflict between Gazprom and the EU Commission (see above), President Putin announced Russia and Turkey's joint intention to construct a new pipeline that would bypass the Ukrainian gas transportation infrastructure and realise Turkey's long-standing dream of becoming an energy hub. In contrast to the original intention, the project may only supply the national market in Turkey instead of expanding to Europe. This is due to several factors, some political, some technical and some due to competing pipelines. As said, although a capacity increase has its own costs, due to the large scale economics of cross-border gas pipelines, it would be more economically viable to build the pipeline with spare capacity than construct another pipeline along the same route.

---

[54]Factual information on Blue Stream pipeline is available at: http://www.gazprom.com/about/production/projects/pipelines/active/blue-stream/.

[55]United Nations Convention on the Law of the Sea, 10 December 1982, 1833 UNTS 3 (UNCLOS).

[56]Energy Charter Treaty, 17 December 1994, reprinted in: ILM 34 (1995), 360.

The Turk Stream pipeline will not bring new and additional gas to Turkey, but instead replace the transit volumes it receives through the Western Line which is currently contracted for private shippers. The existing supply contracts between Gazprom and private shippers will be amended to change their delivery points to Kıyıköy in Turkey, instead of the current delivery point, the Malkoçlar gas-measuring station on the Bulgarian-Turkish border situated on the Western Line. Today, these contracts amount to 14 bcm annual capacity and the first line of Turk Stream will have 15.75 bcm annual capacity. Therefore, the contracted volumes will coincide more or less with the original contracts.

Should the second line be built, the capacity will be reserved to Gazprom, and it is not clear whether private shippers will have access to the pipeline. If the planned two lines of the pipeline are commissioned, then the pipeline would increase transit capacity in Turkey, hence serving Turkey's long-standing ambition to become a hub for natural gas supplies to Southern European consumers. The direct link from Russia to Turkey increases Turkey's energy security (lower transit risks than the Western Line) and increases its political leverage.

Around 700 km of the offshore part will lie within Turkish waters and the Turkish EEZ in the Black Sea; the Turkish offshore section will be divided into two parts. For the section of the pipeline that runs from the Turkish and Bulgarian EEZ to the Turkish coastline, with a length of approximately 275 km, the pipeline company, South Stream Transport, will conduct an EIA, which will be developed as per Turkish legislation and will cover local environmental conditions, communities and overall pipeline safety. For the rest of the route, an EIA has already been approved by Turkey in 2014 as part of the scrapped South Stream project (as it traversed through the Turkish EEZ).

As part of the EIA, South Stream Transport will consult Turkish stakeholders, including government experts, NGOs and fishing groups for the new portion of the offshore route and landfall section in the Kıyıköy region.

The Russian part of the offshore line is developed according to an approved EIA under Russian permitting procedures and uses data made available from the Environmental and Social Impact Assessment (ESIA) of the South Stream, which was developed by engaging communities in the Anapa region near the landfall site, along with NGOs and other interest groups.

The IGA between Turkey and Russia was approved and ratified by Turkey in December 2016 while it was ratified by Russia's lower house of parliament, the State Duma, in January 2017. The agreement will stay in force for thirty years subject to further extension. The project negotiations took two years, as Turkey insisted on a price reduction from Gazprom for Turkey's supply, which the Russian giant was not willing to provide. The shooting down of a Russian jet by Turkey has also prolonged the negotiation period, as President Putin suspended the project. Eventually, the financial and political motivations to construct the Turk Stream pipeline outweighed any previous discord.

As per the IGA, the Turk Stream pipeline will use the same entry point as the South Stream, i.e. the gas compressor station in Russkaya on the shores of the Black Sea in Anapa. For the second line, there will be a long-term supply agreement with a take-or-pay contract with Gazprom, and the full capacity of the second line will be reserved for Gazprom. As mentioned above, the second line will be operated by a joint venture of BOTAŞ and Gazprom (50–50%) and third parties can join the Joint Venture if permission is granted from BOTAŞ and Gazprom. The construction of the second line, however, can be cancelled via notification from the Turkish side as per the IGA.

There will be significant tax exemptions, particularly for offshore operations, and construction will be corporate and profit tax-free; gas supply services will also be tax-free. In addition, there will be no stamp and customs duties for the technical equipment coming from both countries. Disputes will be resolved in the courts of Switzerland.

For Russia, the project is a part of its systematic efforts to avoid certain transit routes. Essentially, for the same reasons leading to the planned Nord Stream 2 pipeline, Russia aims to increase Gazprom's non-transit pipeline capacity and redirect volumes that would otherwise flow through Ukraine. In addition to pursuing its transit avoidance objectives, Russia will also increase its market power in Turkey, its second largest consumer after Germany. Turkey's national gas demand is over 45 bcm per annum and is expected to grow rapidly towards 2030. The Turk Stream pipeline gives leverage to Russia against an already contracted 6 bcm of gas from Azerbaijan, potential flows of gas from the Kurdish Regional Government (KRG) in Iraq to Turkey, and current supplies from Iran for which Turkey pays the highest price.

The IGA of the project indicates that the second gas line will be commissioned only if there is both a demand and commitment from European buyers. In this respect, Turk Stream will be in competition with the TANAP project. Turk Stream, while in principle in competition with TANAP, could be connected to TANAP if Turk Stream's second line is constructed. Gas could then flow through the TAP pipeline, which can be expanded by 10 bcm, and offered to shippers under an open-season procedure as per appropriate EU legislation and accommodate Russian gas alongside natural gas from Azerbaijan (which already has 10 bcm of dedicated TAP capacity).

As per the Incremental Capacity amendment to the EU Network Codes, an amount at least equal to 10% of the incremental technical capacity at the concerned interconnection point, shall be set aside and offered no earlier than the annual quarterly capacity auction.[57] This means that BOTAŞ could increase its transit fees and lighten the burden of its sunk investment costs.

Ostensibly, there are not that many alternative export routes for these supplies, as Turkey is surrounded by resource-rich gas exporting countries in Central Asia, the Southern Caucasus, the Middle East, and North Africa. However, as mentioned above, European indigenous gas production is rapidly falling, while 2016 and 2017 supplies from Russia to Europe saw record highs.

The main reasons for the South Stream's failure included the non-compliance of the project with the EU's Third Energy Package requirements, particularly third-party access and unbundling, as well as competition law concerns. A route to Europe through a non-EU country such as Turkey would allow Russia to access the EU market whilst avoiding such restraints as EU energy law would be applicable from the Greek border onwards. However, on July 2019, Gazprom announced that the second line of the pipeline, if constructed would connect to Bulgaria

---

[57]CAM NC, Art. 8(7).

instead and not Greece. In this case, if the existing volumes from the Western Line were to be replaced by Turk Stream, the spare capacity in the pipeline could be used for reverse flow purposes to Bulgaria and Romania, and natural gas may also be transported to Greece via the Bulgaria-Greece Interconnector with limited capacity. This would allow Gazprom to meet its supply commitments under its capacity contracts with Bulgaria and Romania, which will expire in 2030.

However, the 2019 Amendments to the Gas Directive extends the definition of interconnector to *"transmission line which crosses or spans a border between Member States or between Member States and third countries up to the border of Union jurisdiction"*. This would automatically necessitate them to be brought into line with the capacity allocation network code (i.e. entry/exit tariffs, deliveries at hubs and auctioning of capacity). Alternatively, if the Turk Stream extension was to be made via Bulgaria, both the border interconnectors with Turkey (if a new line is constructed or via reverse flows in the Western Line) and Serbia may apply for individual exemption procedures, and the procedures and grounds for seeking extension (discussed in more detail in Chapter 6) are not entirely straightforward (among others, through requiring extensive consultations among all relevant states in the regional market) and cannot thus be guaranteed.

If connected in Greece, Turk Stream, could have connected to the TAP pipeline operated by BP, Statoil, SOCAR, Fluxys, Enagás and the Axpo. Currently, the TAP pipeline has an exemption from the above-mentioned provisions of the Third Energy Package. An exemption is only provided if a pipeline project brings new sources of gas, it does not damage competition and the exception is indispensable for its development. Regardless of the fact that the TAP pipeline complies with the criteria, the Commission awarded an exemption but only for 50% of its capacity. The total capacity of the pipeline is 20 bcm—10 of which are now dedicated to gas from Azerbaijan via TANAP—but the remaining 10 bcm could have been awarded to Gazprom.

Another issue with potential supplies through Turkey to the EU via the second line of Turk Stream is over reliance on Turkey as a transit country, and as stated above the pipeline would carry Russian gas which would not mean diversification but diversion of supplies (from the Ukrainian route). The lesson from Ukraine is that it is a mistake for any country to rely on a single supplier for its energy source, or any place that has the power to curtail energy consumption in a third country. These developments increase risks in the eyes of the investors particularly in Eastern Europe and inhibit potential pipeline developments. Transit countries have the power to influence markets and have a significant price leverage in addition to revenues from transit fees.

There is always the risk of a shift of 'bargaining power', that once the energy infrastructure is built, the transit country could negotiate and insist on higher rent for passage through its territory. However, in principle, if carefully drafted, legal instruments could provide an effective mitigation mechanism. The fact that Turkey is a contracting party to the Energy Charter which, once allowed, prohibits interruption of transit flow, could provide such assurances for the EU.

Transit conflicts do arise. Despite the fact that both Russia (until 2009) and Ukraine had been parties to the ECT, disputes between these two States relating to the price of gas and transit fees have led to interruptions of transit flows toward the EU. This shows that the existence of legal principles alone may not be sufficient in order to prevent transit interruptions. Legal rules must be backed up by procedural mechanisms under IGAs aimed at the prevention, mitigation and responsibility for energy flows disruptions.

5.5.2 Trans-Caspian Pipeline

The Trans-Caspian pipeline project is considered a necessary link needed to bring Turkmen gas into Europe. In conception since the late 1990s,[58] the project was first suggested in 1996 by the US. In 1997, Turkey and Turkmenistan signed the first agreement on the import of 30 bcm of Turkmen natural gas to Turkey. In February 1999, the Turkmen government agreed with General Electric and Bechtel Group for a feasibility study on the pipeline. At a 1999 OSCE meeting in

---

[58]See more at: http://www.energia.gr/article_en.asp?art_id=30644.

Istanbul, Georgia, Azerbaijan, and Turkmenistan signed a number of agreements concerning the construction of several projects that included the Baku-Tbilisi-Ceyhan and Trans-Caspian pipelines. In the 2010s, the Caspian Development Corporation[59] concept was designed to allow European gas entities to purchase gas from Turkmenistan. The Commission, the EIB, the Public-Private Infrastructure Advisory Facility and the World Bank commissioned a study to examine business models for infrastructure development, gas sales and purchase agreements. One proposal was particularly favoured, which recommended the bundling of gas purchase agreements by multiple stakeholders (joint purchasing arrangements) under long-term ship-or-pay shipping agreements for new pipelines leading to Europe. This project was to assure Turkmenistan and Central Asia in general that Europe is willing to buy large volumes of natural gas, thereby providing signals for upstream developments to accelerate. However, it did not go ahead due partly to the unresolved status of the Caspian Sea and Turkmenistan's unwillingness to open its upstream sector to international oil companies (hence the gas fields remained underdeveloped).[60] There were also capacity concerns as the country's largest gas reserves were in the east, such as the Galkynysh field, whereas the only viable outlet for Europe was located to the west via the Caspian Sea. Therefore, Turkmenistan initially directed its gas flows to China via Uzbekistan and Kazakhstan.

As of 2016, Turkmenistan's proven natural gas reserves are around 17.5 trillion cubic metres.[61] Considering the magnitude of

---

[59]Caspian Development Corporation, Final Implementation Report, December 2010, available at: http://documents.worldbank.org/curated/en/824621468252629798/pdf/689520ESW0P1160mplementation0Report.pdf.

[60]News Article, Reuters, Turkmenistan Boosts Gas Export Capacity with East-West Link, 23.12.2015, available at: https://www.reuters.com/article/turkmenistan-pipeline/turkmenistan-boosts-gas-export-capacity-with-east-west-link-idUSL8N14C0GT20151223.

[61]BP Statistical Review of World Energy, June 2017, available at: https://www.bp.com/content/dam/bp/en/corporate/pdf/energy-economics/statistical-review-2017/bp-statistical-review-of-world-energy-2017-natural-gas.pdf.

Turkmenistan's natural gas reserves, one could still expect Turkmenistan to be able to supply Europe with gas, alongside its Eastern export route. The 2014 Ukrainian political crisis prompted the EU to revert to the Trans-Caspian Pipeline project.[62] In the same year, Turkey and Turkmenistan signed a framework agreement (not disclosed to the public) enabling the supply of Turkmen gas via the TANAP project.[63] Under this agreement, Turkmengas and the private Turkish firm Atagas would purchase Turkmen gas for TANAP. This was followed by the signing of the Ashgabat Declaration on the development of cooperation in the energy field between Turkmenistan, the Republic of Azerbaijan, the Republic of Turkey and the EU.[64]

In 2015, Turkmenistan completed an East-West pipeline connecting Galkynysh and other fields in the east of the country to its Caspian coast, opening up the possibility of Turkmen gas deliveries to Azerbaijan, then to Turkey and on to the EU. Maros Šefčovič implied in 2015 that the plan for Europe is to build all pipelines to its borders by 2020.[65] The Trans-Caspian Pipeline was also backed by the World Bank. In 2015, a joint working group on a project was formed by the signatories of the Ashgabat Declaration. As mentioned above, the authorities signed a Convention in June 2018 on the status of the Caspian Sea. The Convention establishes a formula for dividing up the resources of the Caspian Sea and preventing other powers from

---

[62]Simone Tagliapietra and Georg Zachmann, Designing a New EU–Turkey Strategic Gas Partnership, Bruegel, 01.07.2015, available at: http://www.bruegel.org/nc/blog/detail/article/1666-designing-a-new-eu-turkey-strategic-gas-partnership/.

[63]News release, Reuters, Turkmenistan Inks Deal with Turkey to Supply Gas to TANAP Pipeline, 07.11.2014, available at: https://www.reuters.com/article/turkmenistan-turkey-tanap/turkmenistan-inks-deal-with-turkey-to-supply-gas-to-tanap-pipeline-idUSL6N0SX2QK20141107.

[64]Declaration on the development of cooperation in the field of energy between Turkmenistan, the Republic of Azerbaijan, the Republic of Turkey and the European Union (Ashgabat Declaration), 01.05.2015, available at: https://ec.europa.eu/commission/commissioners/2014-2019/sefcovic/announcements/ashgabat-declaration_en.

[65]News Article, The Diplomat, Europe and Turkmenistan Make Nice: The completion of the East-West Pipeline Could Allow Turkmen Gas to Be Delivered to Europe in the Near Future, 17.05.2015, available at: http://thediplomat.com/2015/05/europe-and-turkmenistan-make-nice/.

establishing a military presence.[66] The agreement will enable construction of an underwater pipeline, such as the Trans-Caspian pipeline and offshore extraction activities in line with international environmental standards.[67] In anticipation of these developments, and with the hope of making the initial Southern Corridor plan a reality, the EU announced in November 2017 that it had included the Trans-Caspian Pipeline project in its PCI list, which receives financing of €1.872 million (50% of the total eligible costs) for a new feasibility study during April-March 2019.[68] The studies involve the establishment of baseline data, engineering and preliminary routing and definition and the award of the proposed route reconnaissance survey contracts, followed up by the design and development of the mechanical design and a reliable estimation of materials and costs needed for a successful PCI development. The status of the action plan is still ongoing.[69] Maros Šefčovič even set a date for Turkmen gas to reach Europe by 2019.[70]

For this project to be realised, a Trans-Caspian pipeline will need to be constructed. The pipeline is expected to be 300 km long, linking with the South Caucasus pipeline, which was expanded in 2017 to enable gas destined for the TANAP and TAP pipelines. Currently, the capacity of the South Caucasus expansion is over 20 bcm per year.[71] Through this route in Georgia the pipeline will split into two: one line will connect to the TANAP pipeline and a second line will feed into a proposed White Stream Pipeline under the Black Sea, connecting

---

[66]News release, BBC, Caspian Sea: Five Countries Sign Deal to End Dispute, 12.08.2018, available at: https://www.bbc.co.uk/news/world-45162282.

[67]News release, Trend, Russian Government Approves Draft Convention on Caspian Sea Status, 22.06.2018, available at: https://en.trend.az/business/economy/2920196.html.

[68]The list of the projects of common interest (PCIs) by country—the (third) Union list of PCIs, 23.10.2017, available at: https://ec.europa.eu/energy/sites/ener/files/documents/member-statespci_list_2017.pdf.

[69]Connecting Europe Facility, Supported Actions—May 2019, available at: https://ec.europa.eu/inea/sites/inea/files/cefpub/cef_energy_brochure_2019-web.pdf.

[70]News Article, The Diplomat, Europe Could Be Getting Turkmen Gas by 2020, 05.05.2015, available at: https://thediplomat.com/2015/05/europe-could-be-getting-turkmen-gas-by-2020/.

[71]More information on South Caucasus Pipeline is available at: https://www.bp.com/en_az/caspian/operationsprojects/pipelines/SCP.html.

Georgia and Romania[72] The White Stream company, based in Georgia, was the recipient of a feasibility study award.[73] Its projected capacity is 30 bcm per year, at an estimated cost of US$5 billion.

On the one hand, one might think that this may not happen in the current demand environment and due to EU policies on reducing dependence on fossil fuels in light of the Paris Agreement.[74] However, only in late 2017, an EU Spokesperson for Energy Union projects revealed that the EU is negotiating with Azerbaijan and Turkmenistan over a Trans-Caspian Pipeline system with an aim to use the Southern Corridor to bring *"gas to the EU from the Caspian Basin, Central Asia, including Iran, the Middle East, and the Eastern Mediterranean Basin"*.[75] On the other hand, Europe is an attractive market for Turkmenistan, which is interested in the diversification of its export outlets. However, traditionally, Turkmenistan would like to abstain from taking any market and transit risks and would instead require purchasers to carry all the risks and costs. Turkmenistan also has other market outlets—consumption centres within the vicinity of the region. One of those is China. Through the Central Asia-China gas pipeline, Turkmenistan exports more than 30 bcm of gas annually to China (as of 2017). The pipeline, which has an annual 55 bcm capacity, also carries Kazakh and Uzbek deliveries along the pipeline corridor and is currently being expanded with a third line to accommodate gas from Kyrgyzstan andIran's Gas Export to Tajikistan.[76]

---

[72]Presentation by Giorgi Vashakmadze, W-Stream Consortium—the promoter of the Trans-Caspian and White Stream Gas Pipeline projects on Trans-Caspian Gas Pipeline—the Project of Common Interest No. 7.1.1 at the Transit Expert Meeting in Beijing on November 2015, organised by the author while working at the Energy Charter Secretariat, available at: https://energycharter.org/fileadmin/DocumentsMedia/Events/20151127-S6_Giorgi_Vashakmadze.pdf

[73]News Article, New Europe, The "Expanded" Southern Gas Corridor: What Comes After 2020? 14.03.2018, available at: https://www.neweurope.eu/article/expanded-southern-gas-corridor-comes-2020/.

[74]News Article, Oil Price, Gazprom Putting the Squeeze on Turkmenistan, 28.07.2015, available at: http://oilprice.com/Energy/Gas-Prices/Gazprom-Putting-The-Squeeze-On-Turkmenistan.html.

[75]News Article, *AzerNews*, Europe Still Hopeful to Get Turkmen Gas, 02.10.2017, available at: https://www.azernews.az/oil_and_gas/119726.html.

[76]Belt and Road Initiative, the Central Asian gas pipeline: A visual explainer, available at: http://multimedia.scmp.com/news/china/article/One-Belt-One-Road/gasPipeline.html.

Furthermore, a Turkmenistan-Afghanistan-Pakistan-India (TAPI) gas pipeline is in the process of construction; the opening ceremony took place in February 2018 in Afghanistan.[77] In total, the pipeline is expected to stretch 1814 km with a total capacity of 33 bcm per year. The project is expected to cost around US$10 billion and construction is due to be completed in 2019, despite security concerns in the region such as the civil war in Afghanistan.[78]

In view of Turkmenistan's gas export plans, realisation of the Trans-Caspian pipeline project in the near term will depend on European buyers' appetite for risk-taking, as Turkmenistan secured several alternative diversified market outlets for its gas to flow into. The commercial viability of the project will be determined by the price, demand, political and regulatory approach towards the natural gas sector with the EU. If the project proves uncommercial, the EU might subsidise the costs by tapping into its strategic funds. In this event, the question arises as to who would be the winners or losers of the project. It may or may not be the European citizens whose taxes are used to subsidise an uncommercial project—or the investment markets may bear the costs.

5.6   Future of Gas from Iran and Iraq to the EU?

As of 2016, the proven natural gas reserves of Iran were estimated as 33.5 tcm,[79] making it the largest natural gas reserve in the world, according to BP's annual review (some statistics place Iran second after the Russian Federation). In the 2015 Energy Union framework there is no explicit reference to Iran. However, there has been steady progress in the EU–Iran energy partnership since the easing of sanctions after a

---

[77]News Article, Reuters, Leaders Launch Start of Afghan Section of TAPI Gas Pipeline, 23.02.2018, available at: https://www.reuters.com/article/us-turkmenistan-afghanistan-gas-pipeline/leaders-launch-start-of-afghan-section-of-tapi-gas-pipeline-idUSKCN1G70PU.

[78]John Roberts, Turkmenistan: The Pursuit of New Markets, Platts, *Energy Economist*, Issue 437, 22.04.2018 (subscription only, copy with the author).

[79]BP Statistical Review of World Energy, June 2017, available at: https://www.bp.com/content/dam/bp/en/corporate/pdf/energy-economics/statistical-review-2017/bp-statistical-review-of-world-energy-2017-natural-gas.pdf.

nuclear deal was reached the under the Joint Comprehensive Plan of Action in 2015.[80] The EU banned the purchase of Iranian oil and gas in 2012. In late April 2017, the Commissioner for Climate Action and Energy, Miguel Arias Cañete, visited Tehran to strengthen energy and climate ties with Iran and opened the first-ever Iran-EU Business Forum on Sustainable Energy.[81]

Prior to the introduction of nuclear sanctions, in 2009 Iran expressed interest in exporting natural gas to Europe as part of the Nabucco project.[82] Previously, in 2008 Turkey and Iran signed a MoU for the Turkish Petroleum Corporation to develop parts of the South Pars gas field. Parties also agreed to the construction of an ITE pipeline across Turkey to deliver natural gas produced in Iran and Turkmenistan to Turkey and Germany (now replaced by Greece and Italy via the Southern Corridor). In 2013, the Turkish Council of Ministers published a Decree for nationalisation of the properties on the proposed pipeline route. In the same year the project developer, a Turkish company (Turang Tasimacilik) granted an incentive certificate worth approximately €5 billion for the construction of the pipeline. The proposed pipeline capacity is 35 bcm a year with a 5000 km pipeline. Due to the sanctions imposed on Iran by Western nations, the completion of this pipeline was postponed, and only the Tabriz-Ankara section remains operational.

The Tabriz-Iran pipeline is the connected Eastern Anatolia Natural Gas Main Transmission Line for transporting natural gas produced mainly in Iran and other countries to the east of Turkey. The 1491 km line was put into operation in 2001 and has an annual 14 bcm maximum through-put capacity. There were two alternatives for this pipeline to supply Europe with gas. The first was to increase the capacity of

---

[80]EU, An EU Strategy for Relations with Iran After the Nuclear Deal, June 2016, available at: http://www.europarl.europa.eu/RegData/etudes/IDAN/2016/578005/EXPO_IDA(2016)578005_EN.pdf.

[81]EU Commission News, Commissioner Arias Cañete in Iran for the first-ever Iran-EU Business Forum on Sustainable Energy, 28.04.2017, available at: https://ec.europa.eu/energy/en/news/commissioner-arias-ca%C3%B1ete-iran-first-ever-iran-eu-business-forum-sustainable-energy.

[82]News Article, RT, Iran looks to join Nabucco axis, 10.10.2009, available at: https://www.rt.com/business/iran-looks-join-nabucco/.

the Tabriz-Ankara route and use the TANAP/TAP pipelines; the second was to construct a pipeline following the Nabucco-West route through Bulgaria and Romania.[83] According to information revealed on the project's website, the Turkish section of the ITE Natural Gas Pipeline will begin at the border of Turkey and Iran and terminate at the Greek border.[84] In 2015, the Chairman of the Iranian National Gas Company for International Affairs indicated that Iran might export natural gas in the volume range of 25–35 bcm to European countries.[85] There were even talks that Naftiran Intertrade Company Sàrl (NICO), a Swiss-based subsidiary of the National Iranian Oil Company (NIOC), could have joined the Southern gas corridor and TANAP following the lifting of sanctions.[86]

Notwithstanding the possibility of exporting gas to the EU, Iran aspires to be the third largest gas producer in the world with gas production of 360 bcm/y by 2025.[87] This requires considerable investment to expand and upgrade its energy infrastructure, which may not happen very swiftly in this low commodity price cycle in existence since mid-2014. Iran gas reserves do not currently represent a viable alternative to Russian supplies to the EU. However, if substantial investment could be attracted to its South Pars field and to expansion of the Southern Gas Corridor, there would could be an opportunity to connect Iranian energy infrastructure

---

[83]Changing pipelines, shifting strategies: Gas in south-eastern Europe, and the implications for Ukraine, July 2015, available at: http://www.europarl.europa.eu/RegData/etudes/IDAN/2015/549053/EXPO_IDA(2015)549053_EN.pdf.

[84]More information on Iran–Turkey–Europe (ITRE) pipeline project is available at: http://en.turangtransit.com.tr/overview.

[85]News Article, APA, Azizullah Ramazani: "Iran seeks ways to export gas to Europe within cooperation with Turkmenistan and Azerbaijan" 19.02.2015, available at: http://en.apa.az/xeber_azizullah_ramazani_____iran_seeks_ways_to__221872.html.

[86]News Article, Natural Gas Europe, Turkey Wants to Increase Iranian Gas Export and Iran Is Keen to Have a Share in TANAP, 09.04.2015, available at: http://www.naturalgaseurope.com/turkey-iran-natural-gas-23086?utm_source=Natural+Gas+Europe+Newsletter&utm_campaign=8d43a0438e-RSS_EMAIL_CAMPAIGN&utm_medium=email&utm_term=0_c95c702d4c-8d43a0438e-307768681.

[87]EU, An EU Strategy for Relations with Iran After the Nuclear Deal, June 2016, available at: http://www.europarl.europa.eu/RegData/etudes/IDAN/2016/578005/EXPO_IDA(2016)578005_EN.pdf.

with Europe.[88] In addition to the low-price environment, the upstream regulatory environment in Iran has traditionally been considered unfavourable by international oil companies (the buyback contracts), and domestic natural gas demand is already very high in Iran and is set to increase. In 2017, the buy-back model was replaced with the Iranian Petroleum Model Contract, and in 2018 many deals were expected to be signed with IOC and NOCs including Shell and CNPC.[89]

Further, the energy partnership between Turkey and Iran has not been trouble-free. The two countries have been involved in an international arbitral dispute over natural gas prices since 2012. In 2016, the court ruled in favour of Turkey and ordered Iran to repay the latter 13% of the income it collected in gas exports since 2011 as compensation.[90] There is also some competition between Azerbaijan and Iran, both having ambitions to supply natural gas to the EU. On the point of integrating Iranian gas into the European system, Šefčovič commented in 2015 that "*It takes time, because most of the gas areas, as far as we know, are more to the south, and of course Europe is more to the north. So it's quite a big distance*".[91] As for the preferred route, he noted that "*it's quite clear that when it comes to gas, the necessary infrastructure will have to be developed, and they (producers) have to decide whether they go for LNG, or for pipeline construction*".[92] However, the managing director of the National Iranian Gas Export Company commented that construction of natural gas pipelines to the EU would not be economically justifiable, as Iran could export gas to neighbouring

---

[88]News Article, Natural Gas Europe, The European Council on Foreign Relations looks at the proposed energy union and all the alternative methods of gas for Europe away from Russian supply, 11.04.2015, available at: http://www.naturalgaseurope.com/analysis-europe-alternatives-russian-gas-23151.

[89]News Article, Presstv, Iran to Sign First Major Post-sanction Oil Deals This Month, 03.03.2018, available at: http://www.presstv.com/DetailFr/2018/03/03/554192/Iran-oil-deals-IPC-North-Azadegan.

[90]News Article, Anadolu Agency, Turkey to Receive $1.9B in Compensation from Iran Over Natural Gas Price Dispute, 27.01.2017, available at: https://www.dailysabah.com/energy/2017/01/24/turkey-to-receive-19b-in-compensation-from-iran-over-natural-gas-price-dispute.

[91]News Article, Politico, the coming Russia–Iran energy axis Russia tries to gain a foothold, not a competitor, 20.07.2015, available at: http://www.politico.eu/article/iran-russia-gas-lng-gazprom-energy/.

[92]News Article, Politico, the coming Russia–Iran energy axis Russia tries to gain a foothold, not a competitor, 20.07.2015, available at: http://www.politico.eu/article/iran-russia-gas-lng-gazprom-energy/.

countries and India (via the proposed TAPI pipeline) and also construct LNG facilities (which may be transported to the EU as well).[93]

Numerous LNG plant construction contracts between Iran and Shell, Total and Repsol, for instance, have been put on hold or abandoned due to the sanctions. Another factor is the utilisation of considerable volumes of gas for re-injection in old oil fields in Iran. Furthermore, a pipeline from the South-Pars field via Turkey to the EU is expected to cost a minimum US$16 billion; and even if constructed, this gas will have to compete with possible gas supplies in the same direction from the Iraqi Kurdish region.[94] The key question here is who will finance these investments, in light of the demand and regulatory uncertainty in the EU and low oil and gas prices worldwide. Price cyclicality is difficult to predict. The oil price has already recovered in 2018 from its slump since mid-2014, and although the appetite for major LNG pipeline projects has diminished, one could still consider Iranian natural gas feeding into the TANAP pipeline as the most economically viable option. However, the US has withdrawn from the Iran nuclear deal on May 2018,[95] and tensions are accelerating with the US re-imposing the sanctions that had previously been lifted or waived under the nuclear deal. In 2019, the situation is that the two states are at the brink of war, and the situation does not seem likely to be settled in the near future. Therefore, it is fair to conclude that future of Iranian gas in the EU energy mix is not foreseeable.

The possibility of Iraqi exports of natural gas to Europe has also been on the agenda since the Nabucco project was first designed. The Nabucco project was abandoned, but the EU still has not given up on energy cooperation with Iraq. The EU actions for expanding the

---

[93]News Article, Presstv, Iran Says Piping Gas to EU Not Economical, 01.08.2105, available at: http://www.presstv.ir/Detail/2015/08/01/422835/Iran-Gas-Pipeline-Europe-Not-Economical.

[94]News Article, Natural Gas Europe, Iran's Gas Export to the EU: When, How and How Much? 22.09.2015, available at: http://www.naturalgaseurope.com/iran-gas-export-to-eu-wood-mackenzie-analysts-25479?utm_source=Natural+Gas+Europe+Newsletter&utm_campaign=de36a26196-RSS_EMAIL_CAMPAIGN&utm_medium=email&utm_term=0_c95c702d4c-de36a26196-307768681.

[95]News article, Financial Times, What the US Withdrawal from the Iran Nuclear Deal Means, 09.05.2018, available at: https://www.ft.com/content/e7e53c72-538c-11e8-b3ee-41e0209208ec.

Southern Gas Corridor encompass working closely with gas suppliers in the region, including Iraq.[96] In the Communication on 'Energy infrastructure priorities for 2020 and beyond', the Commission outlined a workplan for an integrated establishment of key interconnections with third countries, in view of falling gas production in the EU. This document also presented the Commission's willingness to help suppliers in neighbouring regions to develop their energy sectors in an efficient and sustainable way; in other words, to enable them to open up their markets in line with the principles and investment patterns of the EU.[97]

The majority of oil and gas reserves in Iraq are concentrated in the Shiite areas of the south and the ethnically Kurdish region in the north, and some reserves are under the control of the Sunni minority in central/western Iraq. According to data from the BP energy outlook report, by the end of 2016, Iraq had an estimated 3.7 tcm of total proven natural gas reserves.[98] The website of the Ministry of Natural Resources of the KRG suggests that there are 5.67 tcm of recoverable natural gas in the Kurdish Region bordering the Eurasian energy axis.[99]

The KRG emerged recently as an important player in supplying Turkey with natural gas. Turkey and the KRG have in fact signed a natural-gas sales agreement in 2013 for the export of 4 bcm/y by 2017, 10 bcm/y by 2020 and 20 bcm/y thereafter.[100] The Turkish-British Company Genel

---

[96]More information on Southern Gas initiative is available via: https://ec.europa.eu/energy/en/topics/imports-and-secure-supplies/gas-and-oil-supply-routes.

[97]Communication from the Commission to the European Parliament, the Council, the European Economic and Social Committee and the Committee of the Regions on security of energy supply and international cooperation—"The EU Energy Policy: Engaging with Partners Beyond Our Borders," 7.9.2011, COM (2011) 539 final.

[98]BP Statistical Review of World Energy, June 2017, available at: https://www.bp.com/content/dam/bp/en/corporate/pdf/energy-economics/statistical-review-2017/bp-statistical-review-of-world-energy-2017-natural-gas.pdf.

[99]More information on Kurdish Regional Government's natural gas reserves and vision is available at: http://mnr.krg.org/index.php/en/component/content/?view=featured.

[100]News Article, *Hurriyet Daily*, KRG Plans 10 bcm in Natural Gas Exports to Turkey in Two Years, 20.11.2015, available at: http://www.hurriyetdailynews.com/krg-plans-10-bcm-in-natural-gas-exports-to-turkey-in-two-years-91471; SIMONE TAGLIAPIETRA and GEORG ZACHMANN, Designing a New EU–Turkey Strategic Gas Partnership, Bruegel, 01.07.2015, available at: http://www.bruegel.org/nc/blog/detail/article/1666-designing-a-new-eu-turkey-strategic-gas-partnership/.

Energy has been developing the Miran and Bina Bawi fields.[101] Further, Turkish authorities have often referred to natural gas from KRG as the cheapest option for imports to Turkey.[102]

However, there are several obstacles to cross-border natural gas infrastructure development from KRG to Turkey and on to Europe for export. The first is obvious: it is the infrastructure security risks. To overcome this, KRG has made significant investments in security of energy infrastructure in recent years, and the region has largely defeated threats from terrorist groups including the Islamic State of Iraq and the Levant. Second, the political relationship between Turkey and KRG are turbulent due to strong opposition from the Turkish Government to the possible independence of KRG.[103] Third, there is a growing demand for natural gas in Iraq predominantly for electrification and tackling energy poverty. In 2016, KRG received a US$375 million loan from the World Bank to increase electricity generation from 1000 megawatts to 1500 megawatts.[104] Finally, Iraq and the KRG need to attract investment to develop their upstream gas production, and also for export infrastructure and transit capacities to be constructed.

All in all, it is not likely, in the short and medium term, to expect Iraqi natural gas to flow into Europe. In the long term, the role of natural gas in the EU energy mix is highly uncertain, and as mentioned in the previous chapters of this book, after 2050 natural gas will be high carbon, and its role should ideally be replaced by renewable energy sources and hydrogen. However, in 2017 Russia's Rosneft agreed to

---

[101]Press release, Genel Energy plc, Bina Bawi and Miran West gas resource update, 19.01.2018, available at: http://genelenergy.com/media/2172/genel-energy-miran-and-bina-bawi-gas-resource-update-final.pdf.

[102]Gokce Mete, Turk Stream Pipeline Project: An Analysis of Legal, Financial and Technical Aspects Publication, European Centre for Energy and Resource Security, Reflections, Working Paper Series, Volume 3, Spring 2017, available at: https://www.kcl.ac.uk/sspp/departments/warstudies/research/groups/eucers/pubs/reflections-3.pdf.

[103]Gulmira Rzayeva, Gas Supply Changes in Turkey, January 2018, Oxford Institute of Energy Studies, available at: https://www.oxfordenergy.org/wpcms/wp-content/uploads/2018/01/Gas-Supply-Changes-in-Turkey-Insight-24.pdf.

[104]News release, IFC, Rebuilding Iraq's Energy Infrastructure, May 2016, available at: https://www.ifc.org/wps/wcm/connect/news_ext_content/ifc_external_corporate_site/news+and+events/news/rebuilding+iraqs+energy+infrastructure.

fund a natural gas pipeline in KRG by investing more than US$1 billion. The estimated pipeline capacity is 30 bcm with possible gas exports, which could correspond to 6% of EU total gas demand. It is reported that pipeline construction will begin in 2019 and export could start as early as 2020.[105] This project and Rosneft's acquisition of 30% of Egypt's Zohr gas field in the Mediterranean Sea in 2017,[106] could be seen as an effort to break the export monopoly of rival Gazprom to the EU. While production from the Zohr gas field has reportedly more than doubled in 2019, pipeline development in the KRG for natural gas exports may not be completed anytime soon. Many factors will play in the realisation of this project; in the current structure of EU decision-making in the energy sector, a Russian financed natural gas corridor may not be well perceived by the EU authorities, as the case of North Stream 2 has shown.

5.7   Will There Be Imports from East-Med?

Significant natural gas discoveries have been made in Israel. The first gas deposit was discovered in 2009 at the Tamar field located 90 km west of the port city of Haifa. The Tamar field has an estimated 240 bcm of recoverable gas reserves. The second discovery, two years later, is the Leviathan field, located in deep water 30 km west of Tamar with studies concluding 450 bcm of recoverable gas. Therefore, Israel has emerged as an important natural gas player—not only in the Middle East but also for Europe. Further discoveries were made between 2011 and 2013 in Israel in the Karish, Tanin, Dolphin, Tamar SW, Aphrodita-Ishai and Shimshon fields. The total proven reserves in Israel in 2016 was 200

---

[105]Press release, Kurdish Regional Government Ministry of Natural Resources, KRG and Rosneft Deal on Construction of Natural Gas Pipeline, Exports Expected in 2020, 18.09.2017, available at: http://mnr.krg.org/index.php/en/press-releases/596-krg-and-rosneft-deal-on-construction-of-natural-gas-pipeline,-exports-expected-in-2020.

[106]Press release, Rosneft, 09.10.2017, available at: https://www.rosneft.com/press/releases/item/188045/.

bcm,[107] according to BP. The development of these fields in Israel is an ongoing economic activity with the first gas deliveries expected by the end of 2019.[108]

In 2011, Noble Energy discovered the Aphrodite Field offshore of Cyprus which is reported to hold 120 bcm reserves.[109] In February 2018, Eni and Total announced that they had discovered a promising natural gas field in the EEZ of Cyprus.[110] But the question is: where are these potential gas supplies destined to go? No doubt these discoveries will contribute to the economies of Cyprus[111] and Israel, as they will reduce dependency on external supplies for both countries and may even make them self-sufficient. Yet, on top of domestic supplies, both of these countries also have resource rent maximisation in mind. Is there a feasible route to market for the so-called East-Med gas?

The most effective route to the EU which has been proposed is a pipeline through Turkey. In 2015, Turkey and Israel started negotiations for the construction of an offshore pipeline through the EEZ of Cyprus to Turkey, which could potentially supply Turkey with 10 bcm of gas and up to 20 bcm for Europe. The planned pipeline would end at the Turkish port of Mersin with a total construction cost of up to US$3 billion.[112] From there, the pipeline could link up with the TANAP pipeline in north-east Turkey.

---

[107]BP Statistical Review of World Energy, June 2017, available at: https://www.bp.com/content/dam/bp/en/corporate/pdf/energy-economics/statistical-review-2017/bp-statistical-review-of-world-energy-2017-natural-gas.pdf.

[108]More information on the history of natural gas exploration in Israel EEZ is available at: http://www.energy-sea.gov.il/English-Site/Pages/Oil%20And%20Gas%20in%20Israel/History-of-Oil--Gas-Exploration-and-Production-in-Israel.aspx

[109]More information on Aphrodite Gas Field is available at: http://www.delekdrilling.co.il/en/project/aphrodite-gas-field

[110]News Article, Reuters, Eni/Total Find Natural Gas Off Cyprus in Field Close to Zohr, 08.02.2018, available at: https://www.reuters.com/article/us-cyprus-natgas/eni-total-find-natgas-off-cyprus-in-field-close-to-zohr-idUSKBN1FS1G3.

[111]In the case of Cyprus, domestic use of natural gas will require natural gas transport and distribution networks, gas power plants and industrial facilities that can use gas to be constructed, as there is currently no natural gas infrastructure in the country.

[112]News Article, Natural Gas Europe, Turkey Plans to Build a Pipeline From Israel, 20.12.2015, available at: http://www.naturalgaseurope.com/turkey-plans-to-build-a-pipeline-from-israel-27244?utm_source=Natural+Gas+Europe+Newsletter&utm_campaign=b614fcd5a7-RSS_EMAIL_CAMPAIGN&utm_medium=email&utm_term=0_c95c702d4c-b614fcd5a7-307768681.

In 2017, these pipeline talks were still ongoing and at the 22nd World Petroleum Summit in Turkey that summer, the adviser to Israel's National Infrastructure, Energy and Water Resources Minister, Yuval Steinitz, revealed that discussions on the price and the proposed route between the two countries have already started. In addition, Yuval Steinitz indicated that Turkey and Israel have agreed to try to conclude the inter-governmental umbrella agreement that will enable the construction of the Turkey-Israel pipeline by the end of 2017.[113] However, US President Trump's decision to move the US embassy to Jerusalem caused a deterioration of political ties between Turkey and Israel.[114]

The European direction also has an alternative route, called the East-Med pipeline project, which has been selected as a PCI.[115] This pipeline aims to connect the Leviathan gas field in the Mediterranean Sea off the coast of Israel, to Italy and Greece via southern Cyprus and Crete. The project received CEF funding for its preliminary front end engineering and design (FEED) studies during 2015–2018 which included economic, financial and competitiveness studies alongside technical feasibility. It reported that *"taking into account the need of additional net imports to satisfy EU gas demand by 2030 and the risk associated to the current production availability, procurement and transport of gas supply, the Project provides strategic contribution to the EU security of supply"*.[116] Furthermore, the outcome of the studies concluded that the project is

---

[113]News Article, *Hurriyet Daily*, Turkey, Israel, Discuss Price, Route of Gas Pipeline, 13.09.2017, available at: http://www.hurriyetdailynews.com/turkey-israel-discuss-price-route-of-gas-pipeline-120807.

[114]News Article, *Hurriyet Daily*, Erdoğan warns Trump: Jerusalem is Muslims, Red Line, 05.12.2017, available at: http://www.hurriyetdailynews.com/erdogan-warns-trump-jerusalem-is-muslims-red-line-123579.

[115]Pipeline from the East Mediterranean gas reserves to Greece mainland via Crete with metering and regulating station at Megalopoli (7.3.1). The 2017 list of PCIs are available here, https://ec.europa.eu/energy/sites/ener/files/documents/annex_to_pci_list_final_2017_en.pdf.

[116]CEF, Eastern Mediterranean Natural Gas Pipeline—Pre-FEED Studies, Part of PCI No. 7.3.1, executive summary available at: https://ec.europa.eu/inea/sites/inea/files/cefpub/summary_7.3.1-0025-elcy-s-m-15_final.pdf.

technically feasible, economically viable and commercially competitive and it would diversify supply sources for Europe and reduce the end consumers' gas bills. The FEED stage of the project therefore has already started.

In March 2018, the Israeli Energy Minister announced that a decision to build the 2000 km East-Med pipeline will be made by early 2019.[117] Israel, Cyprus, Greece and Italy are expected to sign an IGA and the Commission included East-Med in 2019 consultations of the PCI selection, with a decision expected in late 2019.

The project is not only backed by the EU but also the US and has an estimated cost of US$7 billion. Despite the outcome of the pre-FEED studies, the economics of a pipeline of this magnitude remain questionable in view of the current gas prices in the EU, demand uncertainty and the small volumes that the pipeline could potentially carry. Without further proven discoveries in Cyprus, this project, if constructed, would not be based on commercial viability but political considerations, as it would take a long time to pay back this investment. Instead, gas deliveries to Turkey, with its ambitious energy policy to become a trading hub, would make more commercial sense, as the cost of the infrastructure would be significantly reduced. Also, Turkey would buy large volumes to diversify its natural gas mix which would in principle make the project a more marketable option for producers.

The pipeline from Turkey would cross the EEZ of Cyprus does not seem feasible in the foreseeable future however due to the ongoing tensions between Cyprus and Turkey, and between Turkey and the EU. Cyprus remains ethnically divided, and Turkey fiercely opposes exploration activities in Cyprus, claiming that Greek Cypriots lack jurisdiction.[118] Furthermore, Turkish drilling activities in the Eastern Mediterranean received serious backlash from the European Council.

---

[117]News Article, Reuters, Israel Expects Decision on East Med Gas Pipeline to Europe in 2019, 08.03.2018, available at: https://uk.reuters.com/article/us-ceraweek-energy-israel/israel-expects-decision-on-east-med-gas-pipeline-to-europe-in-2019-idUKKCN1GK2OI.

[118]News Article, Turkey to Send Drill Ship to Contested Gas Field Off Cyprus: President Erdoğan Says He Will Not Let Natural Reserves Be Exploited by Greek Cypriots, https://www.theguardian.com/world/2018/mar/22/turkey-to-send-drill-ship-to-contested-gas-field-off-cyprus

The latter suspended negotiations on the Comprehensive Air Transport Agreement and agrees not to hold the Association Council and further meetings of the EU-Turkey high-level dialogues for the time being. The Council also endorsed the Commission's proposal to reduce the pre-accession assistance to Turkey for 2020 and invited the EIB to review its lending activities in Turkey, notably with regard to sovereign-backed lending.[119] Increasing reliance on Turkey as a transit state, therefore would not be appealing to the EU.

Cyprus also has the option to construct an LNG terminal and export natural gas to Europe or build a pipeline connected to Egypt's existing LNG terminals. Israel also has other outlets to export its natural gas: the Noble Energy and Delek Group have already signed deals with Jordan in 2014 and 2016 for natural gas deliveries from the Tamar and Leviathan fields.[120] Furthermore, it was announced in February 2018 that Egypt will buy 64 bcm of natural gas from Israel over the course of a decade.[121]

In view of the alternatives available to producers in the East-Med region, Israel and potentially Cyprus, and the economic realities of pipeline construction to Europe circumventing Turkey, it is clear that the decision to go with the East-Med route would not be a commercial decision but a political one. The Turkish route would be commercially more viable but is fraught with uneasy political relationships between Turkey and Israel, and Turkey and the EU, as well as the unresolved Cyprus problem. All in all, the feasibility of the Turkey route would be complicated under the 2019 Gas Directive Amendment, which changed significantly the EU's policy and regulation towards pipelines coming from third countries.

---

[119]European Council Conclusions, 20.06.2019, available at: https://www.consilium.europa.eu/media/39922/20-21-euco-final-conclusions-en.pdf.

[120]News Article, Bloomberg, Fraying Israel–Turkey Ties Threaten Planned Natural Gas Venture, 06.02.2018, available at: https://www.bloomberg.com/news/articles/2018-02-05/fraying-israel-turkey-ties-threaten-planned-natural-gas-venture.

[121]News Article, Reuters, UPDATE 3-Egyptian Firm to Buy $15 Billion of Israeli Natural Gas, 19.02.2018, available at: https://uk.reuters.com/article/israel-egypt-natgas/update-3-egyptian-firm-to-buy-15-billion-of-israeli-natural-gas-idUKL8N1Q92NX.

5.8   Interconnections

5.8.1   Baltic Pipe

The Baltic Pipe project is a planned 900 km direct connection between Norwegian supplies and Central-Eastern Europe and the Baltic region. The capacity of the pipeline is up to 10 bcm and envisaged to have a 3 bcm reverse capacity from Poland to Denmark. Construction is due to begin in 2020 in cooperation between the Polish TSO, GAZ-SYSTEM and the Danish TSO, Energinet.dk.[122]

This project is linked with the Balticconnector pipeline, connecting Finland and Estonia and the Gas Interconnection project between Poland and Lithuania (GIPL), as the purpose is to create a joint Baltic and Finnish gas market with a potential single entry-exit zone. Essentially this will bring gas prices in the Baltic region to the same level as in the North-West Europe markets for the benefit of consumers. Receiving €215 million of CEF funding in 2019, the Baltic Pipe is planned to be fully commissioned by 2022[123]

Alongside this project, Poland's LNG terminal in Świnoujście received subsidies worth €128 million received from the EU ERDF for its capacity extension from 5 to 7.5 bcm.[124] The projects mentioned above aim to break Poland's reliance on Russian supplies as Poland imports around two thirds of its natural gas from Russia although the current contract with Gazprom is due to come to an end in 2022.

---

[122]Baltic Energy Market Interconnection Plan (GRIP) Main Report, 2017, available at: https://www.entsog.eu/sites/default/files/files-old-website/publications/GRIPs/2017/entsog_BEMIP_GRIP_2017_Main_web_s.pdf.

[123]European Union, Press release, €215 Million of CEF Energy Funding for the Baltic Pipe Connecting the Danish and Polish Gas Transmission Systems, 15.04.2019, available at: https://ec.europa.eu/inea/en/news-events/newsroom/%E2%82%AC215-million-cef-energy-funding-baltic-pipe-connecting-danish-and-polish-gas.

[124]European Union, Press release, Extension of Liquefied Natural Gas (LNG) Terminal: Increasing Poland and EU's Energy Security, 24.04.2019, available at: https://ec.europa.eu/regional_policy/en/newsroom/news/2019/04/24-04-2019-extension-of-liquefied-natural-gas-lng-terminal-increasing-poland-and-eu-s-energy-security.

However, these investments are not planned as a result of the real need for a capacity increase or new gas infrastructure (as most LNG plants, and some pipelines are still under-utilised in the EU). The Baltic Pipe and Polish LNG capacity expansion projects are essentially strategic projects that do not necessarily make commercial sense and hence have needed topping up with public resources. They present a business as usual scenario, to push for security of supply infrastructure to import natural gas from elsewhere. They also hold the promise of bringing gas prices down. Saying that, Poland never discloses its import prices for Qatar or Russia, therefore it is not guaranteed that Qatari gas supplied to the Swinoujscie terminal would be cheaper than Russian pipeline supplies.

5.8.2 Bulgaria-Romania-Hungary-Austria (BRUA)

This project is another interconnector that received, €179 million in CEF funding and an additional €100 million from the EFSI through the EIB in 2017. The total estimated cost of this major pipeline infrastructure project traversing Bulgaria, Romania, Hungary and Austria, is around €560 million.[125] The project received a further loan worth €60 million towards completion of the construction of the Romanian section of the gas pipeline in 2019.

The pipeline aims at providing resources from Romania's offshore zone as well as connecting with Turkey as it is one of the two routes of the Southern Corridor. The 1318 km BRUA pipeline is designed to have an annual throughput of 23 bcm. Romania's TSO, Transgaz Medias, has already completed 215 km of pipeline and three gas compression stations in 2019, despite the fact that the Hungarian TSO, FGSZ, abandoned plans to develop its interconnector with Austria in June 2018 as it considered existing capacity at the Slovak border to be sufficient. The project also received a deadlock in the negotiations

---

[125]Trinomics, The role of Trans-European gas infrastructure in the light of the 2050 decarbonisation Targets: Specific Tender under Framework Contract, 24.09.2018, available at: https://ec.europa.eu/energy/sites/ener/files/documents/gas_infrastructure_2050_report_tasks_3_and_4_v2.pdf.

between Romanian government and neighbouring countries and the main offshore operator, OMV-ExxonMobil. This is important as BRUA's commercial feasibility depends on the potential for Romanian gas production which is expected to provide a cheap and indigenous natural gas source for Central and South-Eastern Europe. The pipeline could, in addition to the TAP pipeline, connect to the second leg of a potential Trans-Caspian Pipeline. However, suspension of the pipeline's initial route and uncertainty around key conditions in negotiations among the parties involved makes the future of the pipeline ambiguous to say the least.[126]

The Hungarian decision was based on an open season plan which was carried out only at the Romanian-Hungarian border and not on the Hungary-Slovak border. ACER decided that this connection should be built based on the lower level (75%) of offer proposed, instead of the proposed level of 100% of available capacity to be auctioned, in order to allow sufficient flexibility for handling gas from all potential sources and suppliers.[127] In Hungary, there are two TSOs, one large and the other smaller, with the latter having only the Hungary-Slovak interconnector which is indeed running empty. The Hungarian TSO considered that the capacity of Hungarian-Slovak and Slovak-Austrian connection was sufficient, and that there is no need to construct a connection between Austria and Hungary. As exemplified throughout this book, pipeline projects involving multiple states often involve diverging interest and are high-risk investments. However, if the market decides and there is an additional need for 10 bcm capacity, the Hungary-Austria border interconnection could be constructed but there was no open season procedure. Despite the CEF funding provided, the economic and social value of the project so far remains tentative.

---

[126]News Article, Reuters, Romania's Black Sea Gas Projects Hanging by a Thread, 01.04.2019, available at: https://www.reuters.com/article/us-romania-energy-offshore-analysis/romanias-black-sea-gas-projects-hanging-by-a-thread-idUSKCN1RD2HS.

[127]Decision No. of 05/2019 of ACER of 9 April 2019 on the Incremental Capacity Project Proposal for the Mosonmagyarovar Interconnection Point, available at: https://www.acer.europa.eu/Official_documents/Acts_of_the_Agency/Individual%20decisions/ACER%20Decision%2005-2019%20on%20HUAT.pdf.

5.9   Overview of LNG Infrastructure Projects

In recent years the share of LNG in the EU energy mix has increased, and in response the Commission presented in February 2016 a sustainable energy security package aiming to improve the access of Member States to LNG as an alternative to pipeline gas.[128] This package included a non-legislative EU strategy for LNG and gas storage. This strategy promotes construction of the necessary infrastructure to complete the internal market and targets Member States with only one supply source as a measure to safeguard supply security. The strategy highlights the importance of LNG IGAs and contracts to comply with EU law, and that the Commission will work towards establishing standardisation and a regulatory framework for LNG in shipping.

While the document promotes development of new LNG regasification terminals it also requires a more efficient use of existing capacities before any further investment decisions are made to avoid stranded assets in fossil fuel infrastructure. Indeed, the EU already has significant import capacity which could meet up to 43% of its gas demand in 2015, although the access of Member States in South East Europe, the Baltic Region and South East Europe is limited as most of the regasification capacity is concentrated in the coast line of North West and South West Europe. This is where regional hubs and interconnections play an important role.

In 2018, there were 24 large-scale LNG import terminals in the EU and eight small-scale LNG facilities in Finland, Sweden, Germany, Norway and Gibraltar. Four of the large-scale terminals are floating storage and regasification units (FSRUs), the rest being land based. Malta has a Floating Storage Unit (FSU) and onshore regasification facilities. The EU, based on the 2016 strategy document mentioned above, aims to expand its regasification capacity significantly towards 2022.

---

[128]Communication from the Commission to the European Parliament, the Council, the European Economic and Social Committee and the Committee of the Regions on an EU strategy for liquefied natural gas and gas storage, COM (2016) 49 final, 16.2.2016.

LNG projects, like cross-border pipelines, are capital intensive and long-term investments, therefore their commercial viability is very important. Receiving terminals can be financed through end-user tariffs and reflected on gas consumers' bills or gas companies may provide project or corporate finance in return for the right to use the terminal through long-term capacity booking. However, if the LNG terminal is a strategic, security of supply infrastructure, the business case may not be there (and most LNG terminals have been underutilised in the EU). In this case, once again, public funds through CEF and loans from the EIB under EFSI are used to bridge the financing gap (similar to the PCI gas pipelines). As mentioned above, public funds have been recently used to support the Polish LNG extension and the EU has also approved State Aid for the Klaipeda LNG terminal in Lithuania,[129] provided €101.4 million CEF funding for an LNG terminal on the island of Krk in Croatia,[130] whilst the CyprusGas2EU project received €101 million in EU funding.[131]

Increasing LNG receiving capacity does not automatically secure alternative supplies. In 2017, LNG imports only accounted for around 14% of total gas imports, with the bulk coming from Qatar, Algeria and Nigeria. Attracting further supplies, among others, requires securing upstream investments with long-term commitments from buyers. Gas markets in the EU, however, are increasingly dominated by shorter term contracts and buyers do not prefer long-term rigid contracts. This is a risk in the long term for sustainability as final investment decisions are not being taken upstream. There is a mismatch. Removal of

---

[129]European Commission, Press release, State Aid: Commission Approves Support for Klaipėda LNG Terminal in Lithuania, 31.10.2018, available at: https://europa.eu/rapid/press-release_IP-18-6266_en.htm#targetText=In%20November%202013%2C%20the%20Commission,security%20of%20supply%20in%20Lithuania.

[130]European Union, Press release, €101.4 Million of CEF Energy Funding to Improve Energy Security of Croatia and the Region, 18.12.2017, available at: https://ec.europa.eu/inea/en/news-events/newsroom/101.4-million-cef-energy-funding-to-improve-energy-security-croatia-and-region.

[131]European Union, Press release, CEF Energy: €873 Million EU Investment in Energy Infrastructure Adopted by EU Member States, 25.01.2019, available at: https://ec.europa.eu/inea/en/news-events/newsroom/cef-energy-873-million-eu-investment-energy-infrastructure-adopted-eu-member.

destination clauses, on the one hand, is welcome as natural gas can flow to where it is most needed and valued. On the other hand, this means that if the right market conditions are not in place in the EU, cargoes will not find their way to the West. Therefore, the EU is currently working together with Japan (as together they make up 50% of global LNG imports) to standardise destination diversion clauses to provide more market and price transparency and optionality to buyers in leveraging from the diversion.[132]

LNG is important for the EU's clean energy ambitions as it can also be used as a maritime fuel, and contribute to coal to gas switching, thereby reducing GHG emissions in the short term. It is also an important element of the Energy Union's diversification objective. Gas imports will continue in the short and mid-term as a result of decreasing indigenous production. EU LNG imports may reach 100 bcm by the end of 2019 and this may mean a 120% increase in LNG imports. Furthermore, it was noted in 2019, at an EU and US High-Level Business-to-Business Energy Forum that US LNG export has already increased by 272% in 2018, and traded volumes in 2019 were more than 1.4 bcm.[133] It should be noted that while this represents a significant increase, it is still a small amount taking into account that total gas imports in 2018 were around 100 bcm.

Irrespective of the availability of LNG cargoes, 2021–2027 CEF funding in the next EU budget indicates that public money will continue to be used for new mega-pipelines for gas or LNG terminals, instead of leaving it to the market to decide whether to invest in natural gas infrastructure projects. The private sector may be better positioned to navigate through the risk of stranded assets or investing in the renewable energy sector (and new low carbon technologies such as hydrogen).

---

[132]Workshop Series on Key drivers to promote the liquidity, flexibility and transparency of the global Liquefied Natural Gas market Co-organised by the Ministry of Economy, Trade and Industry of Japan and the European Commission Directorate-General for Energy, 2018, available at: https://www.meti.go.jp/press/2018/01/20190107003/20190107003-1.pdf.

[133]European Commission, Press release, U.S. Liquefied Natural Gas Exports Up by 272% as EU and U.S. Host High-Level Business-to-Business Energy Forum, 02.05.2019, available at: https://europa.eu/rapid/press-release_IP-19-2313_en.htm#targetText=In%20terms%20of%20the%20EU's,up%20from%205%25%20in%202016.

5.10 UNFCCC Framework and Paris Agreement

The EU has been in a leading position in the global fight to tackle climate change, and began developing a climate policy as early as 1990 when the UN Intergovernmental Panel on Climate Change (IPCC) published its first assessment report addressing the climate problem scientifically.[134] Today, with the Paris Agreement, virtually the entire international community has agreed to hold the increase in global average temperature to well below 2 °C above pre-industrial levels and to pursue efforts to limit the temperature increase to 1.5 °C above pre-industrial levels under the Parties Agreement. It is worth pausing to realise that this took almost two decades to achieve since the Kyoto Protocol which set forth specific limits on emissions.

The EU (then the European Community) signed the 1997 Kyoto Protocol[135] and were at the forefront of the 2015 Paris Agreement negotiations. Under the latter, the EU committed to reduce its GHG emissions by a minimum of 40% by 2030 compared to 1990 levels. This is implemented under its wider 2030 climate and energy framework. As mentioned throughout this book, taking the ambition a step further, the EU adopted a strategic long-term vision for a prosperous, modern, competitive and climate-neutral economy by 2050 in November 2018. Furthermore, the Clean Energy Package has also been a landmark milestone in translating the EU's energy sector decarbonisation ambitions set out by the Paris Agreement into legislation. However, the journey towards a sustainable energy system continues to be challenging and requires additional policy developments and consistency among Member States. Achieving 'across the board' decarbonisation policy at the EU level is difficult, as some countries have set more ambitious targets than those set at the EU level while others, such as some of the Central Eastern European countries, continue the use of coal and provide high fossil fuel subsidies. The result is that this makes it questionable how quick a formulation can be found.

---

[134]IPCC First Assessment Report Overview and Policymaker Summaries and 1992 IPCC Supplement are available via: https://www.ipcc.ch/report/climate-change-the-ipcc-1990-and-1992-assessments/.

[135]UN Doc FCCC/CP/1997/7/Add.1, Dec. 10, 1997; 37 ILM 22 (1998).

How does ongoing public investment in natural gas infrastructure in the EU align with the Paris Agreement and the 2050 vision? Is there a risk of failure to invest sufficiently in renewable energy, decarbonised gases and other technologies needed for lowering GHG emissions, such as direct air capture, with the limited public funds available? Where does European gas demand stand in this climate scenario? While a significant natural demand increase is not foreseen before 2030, a decrease is also not projected (despite energy efficiency improvements). Is there a misalignment? Probably there is no major misalignment in the projections: natural gas is a fossil fuel (and a high carbon one after 2050, and presumably after the decline of coal), but no realistic alternatives exist in the short and even medium-term to completely phase it out of the energy system. At the same time, natural gas infrastructure projects that are currently being developed in the EU seem to represent bad value for money and are not commercially feasible, as otherwise why would they all require topping up by EU funds?

An all-electric future energy system, while ideal for reducing GHG emissions significantly, is technically challenging and would be prohibitively expensive. Gas therefore will have to play a role as a seasonal storage carrier and as a valuable energy carrier, for heating, transport and industry. This is important as achieving these climate targets requires full decarbonisation of energy supply in all sectors by 2050 and not just in the relatively 'low hanging fruit' of the power sector.[136] Replacing all fossil fuels with renewably produced electricity is a major challenge: finding the appropriate locations, storage and transport are enormous tasks simply because of the magnitude of energy volumes. Cross-sectoral solutions have to be intensified and this includes sector coupling (see below).

Surely the dilemma of security of supply, environment and affordability—the so-called energy trilemma—is a major challenge for policy makers. But the extended CEF funding for 2021–2027 and the

---

[136]Geoffrey Wood, Fossil Fuels in a Carbon Constrained World, in Geoffrey Wood and Keith Baker (eds.), *The Palgrave Handbook of Managing Fossil Fuels and Energy Transitions* (Palgrave Macmillan, 2019).

PCIs selected on the basis of the TEN-E regulation may not best serve the needs of a modernised energy infrastructure, with climate goals insufficiently represented in the process. Going forward, the new Commission's 10 priorities continue to focus on the diversification strategy for natural gas with implementation of the Southern Corridor, development of resources in the East-Med and promotion of LNG and access to gas storage. Apart from the third-country projects, gas infrastructure projects that continue to be funded by the EU include strengthening of the transmission network between Lithuania and Latvia, between Bulgaria, Romania and Austria, between Finland and Estonia, between Poland and Denmark, and let us not forget the new and incremental LNG import capacities in Poland, Lithuania, Croatia and Spain.

The decision-making framework for the gas sector will be discussed in the next chapter, however it should be reiterated here that all of the natural gas PCI projects are strategic security of supply infrastructure aimed at diversification mainly away from Russia. The implication of GHG emissions related to the production of natural gas are not represented in the selection of these projects, including emissions related to internal industrial processes, including flaring, venting, leaks, refining and energy used for the production and distribution of natural gas. Today, numerous studies show that the gas sector can be decarbonised, and decarbonised in a cost effective manner by 2050, with an increasing share of renewable and other low carbon gases, while at the same time maintaining the EU's industrial competitiveness. Hence, there is a need for swift policy adjustments where renewable gas can valuably complement renewable electricity and other renewable energy. However, there is still need for clarity in the regulatory and policy framework for the potential market for these sources.

Proving that CO_2 neutral gas can enable decarbonisation of energy intensive sectors of the economy, in line with the EU's climate ambitions would allow the EU to also take the lead in exporting the technology and know-how for an equal distribution of the costs and burdens of the energy transition globally. It is worth pointing out that this may not be easy as different types of renewable and decarbonised gases come with the requirement of different infrastructure and different climate implications as the actual emissions also vary between CCS, hydrogen, biomethane and biogas. These all have to be taken into account when carrying out a cost-benefit analysis. Furthermore their development will call for public financial support until a market can be created.

5.11 Hydrogen and Renewable Gas Infrastructure

Today, global investment is still more heavily weighted on natural gas infrastructure than on renewable installations. The IEA predicts that until the 2040s, energy investments will be 70% government driven with only 30% being driven by the market.[137] The design of appropriate policies and incentives for the private sector therefore is incredibly important for meeting common global goals on reducing carbon emissions and securing sustainable energy supplies, as well as providing access to affordable energy sources and mitigating the environmental impacts of energy sector activities. Of this 70%, according to the IEA most renewable and energy efficiency projects are driven by public policy initiatives. This is expected to change with increased technology and know-how transfer of these technologies from advanced economies to developing countries, although some barriers may persist. While this book mainly gave examples of subsidies provided for natural gas infrastructure projects, the support provided to the renewable energy industries in the EU to kick start a functioning market also presents some lessons learned for decarbonisation of the gas grid with renewable and low carbon gases.

This is not to suggest that investment incentives should be urgently and completely scrapped in upstream facilities. With no investment, global oil and gas production is expected to decrease significantly. However, even in an energy transition pathway there will be a need for upstream investment as heating and industrial demand cannot be

---

[137]IEA, World Energy Investment Outlook 2018, available at: https://webstore.iea.org/world-energy-outlook-2018.

replaced with low carbon sources alone at least for now. In the transport sector, hydrogen trucks will be important and this can be powered by renewable electricity and blue hydrogen with CCS. There will also be high-speed electrified trains and this is more impactful than electric vehicles in Europe.

Development of hydrogen infrastructure is closely linked with the topic of sector coupling and the use of power to X technologies, which is a policy of the European Union (initially developed in Germany) to create greater synergies between power and gas systems.[138] This concept is further elaborated in the next section; however, the prospect of building a hybrid energy carrier system is worth mentioning here. This would in effect allow utilising electricity and gas assets efficiently and obtaining improved flexibility and security of supply. Power to X include those synthetic fuels generated from renewable electricity such as green hydrogen, ammonia, methane, methanol, diesel, gasoline, and kerosene (can be deployed across all sectors—such as transport, heating, industry, power generation—and replace conventional fuels from hydrocarbons as the primary energy source and feedstock). The Commission is in favor of these technologies as they can support the energy transition and also help reducing the need for building new electric lines/pipelines thereby helping reduce public concerns about the new infrastructure impacting environment and landscape. Power to X solutions can also offer the opportunity to import larger capacities of renewable energy which can then be transported via hybrid infrastructure and stored in gas storage facilities. These technologies will be part of a diversified gas grid towards 2050 however they too require strong policy support and putting effective pressure on fossil fuel prices. CCS, blue hydrogen and power to X all require a technology neutral policy approach is necessary (i.e. not only based on electrification). Currently, power to X technologies are still more expensive than the price of natural gas and oil

---

[138]European Parliament Study requested by the ITRE Committee, Sector coupling: how can it be enhanced in the EU to foster grid stability and decarbonise? November 2018, available at: http://www.europarl.europa.eu/RegData/etudes/STUD/2018/626091/IPOL_STU(2018)626091_EN.pdf#targetText=Sector%20coupling%20involves%20the%20increased,reduce%20the%20costs%20of%20decarbonisation.

(without taxes) due to the price of offshore wind but substantial costs decrease is expected in the future.

A market needs to be created for green gases, and for this market to mature, there is a need to move from Research and Development (R&D) and subsidy driven solutions to market-driven technology. In other words, we need to adopt a path towards market maturity. In many countries, certain energy consumers/sectors are obliged to use a specific share of energy consumption from renewable energy sources. These obligations follow the commitment of governments to fulfil renewable energy targets. Development and usage of green hydrogen therefore could be integrated into the regulatory framework by crediting the fuels against renewable energy obligations and targets. Quotas should be as technology-neutral as possible. If renewable energy targets are pursued in a technology-neutral manner, this could facilitate the development of power to X technologies and a potential power to X market as early as 2030. Green hydrogen may not be available for every country in the EU, as it requires ample sources of renewable energy, therefore blue hydrogen can be used to narrow the gap in the interim. However, standardisation must take place regarding the quality of different low carbon and renewable gases under, for instance, an international binding sustainability regulation with EU wide support schemes enabling an efficient invective of a temporary nature.

Many Member States are currently looking at their options in the EU. Hungary is importing 89% of its gas, and they have an ambition to phase out nuclear power and lignite by improving the penetration of solar power. They are also exploring shale gas, and the country is at an early stage in terms of biomethane and gas decarbonisation experience. Like elsewhere in the EU, the private sector is not investing currently in renewable gas without subsidies.

In the UK, gas is residual; a shale gas sector is basically dead in the country and coal is declining far more rapidly than anticipated whilst a potential for nuclear power to exist remains on the table with strong UK Government support. How will the UK then achieve an all electricity future? There is a need for gas infrastructure to work differently instead. The shift will probably be from transmission to distribution in the UK and grids will need to be upgraded to handle new gases.

Yet, who is going to pay for the conversion of home appliances for hydrogen? In the UK, a clustered approach may work where existing industry can be combined with distribution in a combined project. However even that may not work everywhere in the UK.

Although the end goal may be similar, each national pathway will be different even within the EU. To avoid subsidisation of uncommercial, politically motivated long-term projects as it is happening currently in the EU in the natural gas sector, there is a need to deliver a competitive market for renewable and low carbon gases, and not to intervene with the market principles once the competitive market is established. Gas infrastructure already exists, it is relatively cheap and, as mentioned above, the example of scaling up renewable production could be used to ensure that renewable and low-carbon gases contribute to keeping the cost of decarbonisation down.

Therefore, green gas may have to be supported as renewable energy sources did so that electrolysis technology could flourish and fast. Furthermore, gas companies should ideally not be barred from owning and operating electricity infrastructure; as mentioned above this sector coupling will be important for decarbonisation of the gas sector and as such will be discussed in more detail in the next chapter.

5.12 Conclusions

In the EU, a sizable section of the economy relies on natural gas and this represents an enormous challenge to decarbonise. This is however inevitable. The natural gas sector is already experiencing significant challenges. Most of the cross-border projects involving third country producers are complex, requiring long-term commitments, and high sunk costs in the absence of uncertainties on future energy mix. They are also complicated due to many external political factors including transit risks. Intra-EU gas project developments are lacking commerciality and are continuously supported by public loans and grants. Access to upstream resources are increasingly limited, indigenous production is falling and the EU is facing competition in the LNG markets from the Asian markets. Looking at the projections, in the short to medium-term, higher volumes of LNG and incremental and new sources of pipeline gas from suppliers in the Caspian and East-Med will be needed, and the new infrastructure developments are considered necessary in Central Eastern and South East Europe to complete the internal market. Saying that, the natural gas sector may find it increasingly difficult to convince stakeholders and the public on the acceptance of natural gas solutions.

The future of and long-term viability of investments ultimately still lie, to a large extend, in the hands of policy makers in the EU. This chapter has indicated that the current public financial flows for energy projects may not be consistent with the pathway towards low GHG emissions and climate-resilient development. For instance, the EIB's recent proposal (subject to approval by its Board in 2019) sends a strong signal regarding its commitment to properly implementing the Paris Agreement. EFSI funds used for major gas infrastructure can be redirected towards decarbonisation of the gas grid, which if successful, can set an international precedent. The ultimate success of this immense task will depend on how electricity and gas will work together, and all the gases, including biogas, hydrogen, synthetic gas and biomethane, and all available technologies will be needed.

It is important however that renewable gas must not threaten the net-zero EU objective. Renewable gas should not have a free pass to the future energy system. There is limited potential; it can help some sectors to decarbonise but it may not be the silver bullet, neither may CCS. The key is to have a sound decision-making framework in place for deciding on the future energy and infrastructure architecture of the EU, one that does not subsidise sectors that otherwise are truly uncommercial and perhaps not even demanded, but instead supports forward looking technologies with economic feasibility and market potential up to the point that subsidies are no longer required. The current decision-making framework will be discussed in the next chapter with a broad overview of the current and future regulatory picture.

6

Decision Making Framework for Natural Gas Projects in the EU and on the Future Role of Gas

6.1 Introduction

This chapter focuses on how the legal framework and decision-making process in the EU could promote de-politicisation of the market rules and decarbonisation of natural gas. It presents a critical and detailed account of the complex map of EU natural gas legislation, starting with a historical description of the gradual introduction of liberalisation with each energy package. The chapter then explains the present gas market architecture within the framework of the Third Energy Package, Gas Target Model and Network Codes.

This chapter continues to also rely on data on demand projections mainly obtained from the Commission's Quarterly Report on European Gas Markets and case examples. Analysis of the natural gas decision making framework concentrates on the impact of the increasing decision-making powers of public bodies on commercial projects (mainly the Commission) under Article 36 of The Third Energy Package, together with the information exchange regulations. The problem lies with the regulation and policy that leads to public funds being spent on gas

© The Author(s) 2020

G. Mete, *Energy Transitions and the Future of Gas in the EU*, Energy, Climate and the Environment, https://doi.org/10.1007/978-3-030-32614-2_6

infrastructure, erosion of market mechanisms and the regulatory uncertainties that comes for instance with the Gas Directive amendment.

Further, chapter also breaks down the issues which might have an impact on energy companies' decisions to invest in gas infrastructure. It argues that the decisions on which natural gas infrastructure projects should go ahead needs to depend on the economic viability of projects and the 'rate of- return' for capital investments (upstream or midstream) and the EU's commitments under the Paris Agreement and the UN Sustainable Development Goals (SDGs), implemented through the 2050 strategy. Finally, before concluding this book, it returns to the planned 2020 Gas Package and provides policy recommendations, including on the future of sector coupling, and explores alternatives pathways to promote incentives for technologies that can enable decarbonisation of the gas grid at scale.

6.2   A Brief History of Energy Regulation in the EU

The EU we know today is a product of an ongoing gradual integration process started with the signing, by six European nations (France, Germany, Italy, Belgium, Luxembourg and the Netherlands), of the first European regional organisational Treaty establishing the European Coal and Steel Community[1] in 1951. Following the end of the Second World War, the idea was to create a common market to prevent future conflicts in Europe, particularly by ensuring equal access to the sources of coal and steel production. This early formation of a common regional market in the energy sector evolved into a customs union with the signing of the Treaty of Rome[2] in 1957, establishing the European Economic Community. In 1958, the European Atomic Energy Community[3] (EAEC or EURATOM), a separate yet

---

[1]Treaty Establishing the European Coal and Steel Community, 18.04.1951, 261 U.N.T.S. 140.
[2]Treaty Establishing the European Economic Community, 25.03.1957, 298 U.N.T.S. 11.
[3]Treaty Establishing the European Atomic Energy Community, 25.03.1957, 298 U.N.T.S. 167.

geographically overlapping legal entity, was established to create a dedicated market for nuclear energy in Europe.

In 1986, the Single European Act[4] introduced a monetary union within Europe. The geographical scope of European regional market integration has subsequently expanded since its creation from a 6 nation community to 28 nations.[5] A number of treaties establishing various EU institutions and their competencies have revised earlier agreements and, in 2009, the most recent transformation of the European Regional Economic Integration Organization (REIO) came into being under the Lisbon Treaty.[6]

The Lisbon Treaty replaced all previous EC Treaties and consolidated them under the Treaty on the Functioning of the European Union (TFEU)[7] and the Treaty on European Union.[8] Furthermore, it incorporated a separate article on energy issues[9] and restructured key European institutions,[10] thus opening a new phase for the EU as an international actor functioning as a single legal entity. Last but not least, the Lisbon Treaty expanded the EU common commercial policy to cover foreign direct investment, granting the Commission the right to negotiate investment treaties with non-EU states on behalf of the Member States.[11] All in all, since the early 1990s, the EU has strengthened the common market via closer energy relations through the development of market oriented regulatory initiatives aimed to depoliticise energy and promote market integration. Regional and economic integration also deepened gradually since the 28 Member States ceded significant parts of their sovereignty by way of delegation of some decision-making powers to tailor-made institutions shared at the Union level.

---

[4]Single European Act, 1987 OJ L169/1.

[5]At the time of writing, the UK remains a full member of the EU and rights and obligations continue to fully apply in and to the UK.

[6]Treaty of Lisbon Amending the Treaty on European Union and the Treaty Establishing the European Community, 13.12.2007, 2007/C 306/01 (Lisbon Treaty).

[7]Consolidated Version of the Treaty on the Functioning of the European Union 2008 OJ C115/47 (TFEU).

[8]Consolidated Version of the Treaty on European Union 2010 OJ C83/01 (TEU).

[9]Article 194 TFEU.

[10]Article 47 TEU.

[11]Article 188 C Lisbon Treaty.

The Union can only act in a policy area if:

i.   The action forms part of the competences conferred upon the EU by the Treaties (principle of conferral);
ii. In the context of competences shared with Member States, the European level is most relevant in order to meet the objectives set by the Treaties (principle of subsidiarity); and
iii. The content and form of the action does not exceed what is necessary to achieve the objectives set by the Treaties (principle of proportionality).

In accordance with Articles 3 and 4 TFEU, a distinction should be made between the exclusive and shared competences of the EU and Member States. At the same time, in Article 6 TFEU, a coordinating, supporting and supplementing function has been proposed. In accordance with Article 4(2) (i) TFEU, the area of 'energy' falls within the shared competence regulated by Article 2(2) TFEU.

Certain energy-related matters, though, are deemed exclusive to the EU under Article 3 TFEU. For instance, the competitive conditions for energy trade within the internal market or the question of tariffs, when third country energy commodities cross an EU border (in other words a common commercial policy when commodities enter the customs union) appear to fall squarely within the exclusive competence of the EU.

Article 194 of the TFEU on energy stipulates:

In the context of the establishment and functioning of the internal market and with regard for the need to preserve and improve the environment, Union policy on energy shall aim, in a spirit of solidarity between Member States, to:
a) ensure the functioning of the energy market;
b) ensure security of energy supply in the Union;
c) promote energy efficiency and energy saving and the development of new and renewable forms of energy;
d) and promote the interconnection of energy networks.

Without prejudice to the application of other provisions of the Treaties, the European Parliament and the Council, acting in accordance with the ordinary legislative procedure, shall establish the measures necessary to achieve the objectives in paragraph 1.

Such measures shall be adopted after consultation of the Economic and Social Committee and the Committee of the Regions. Such measures shall not affect a Member State's right to determine the conditions for exploiting its energy resources, its choice between different energy sources and the general structure of its energy supply.

Article 194 hence defines the areas of exclusive competence of the Union in the energy field. The Commission uses its competence by legislative action. However, its provision on energy rights of the Member States stipulates that measures that significantly affect a Member State's national energy mix and the general structure of its energy supply shall not be enacted in ordinary legislative proceedings, but rather by unanimous decision of the Council.

Although Article 194 makes no direct reference to the external dimension of EU energy policy, when read together with the implied powers doctrine under Article 3(2) TFEU, which provides the authority to adopt international measures that are necessary to achieve the EU's internal objectives, it can be interpreted that the EU Treaties provide the legal foundation for promoting international, or at least a pan-EU, energy framework based on EU energy policy.[12]

In the case of a competence shared by Member States and the Union, the EU legal provisions on the matter are conclusive if the Union has chosen to exercise its competence. A Member State accordingly has regulatory authority only when the Union has issued no regulations or has retracted a previously existing regulation.[13]

---

[12]Peter Van Elsuwege, The EU's Governance of External Energy Relations: The Challenges of a Rule-Based Market Approach, in D. Kochenov and F. Amtenbrink (eds.), *The European Union's Shaping of the International Legal Order* (Cambridge University Press, 2013), 218.

[13]Ulrich Ehricke and Daniel Hackländer, European Energy Policy on the Basis of the New Provisions in the Treaty of Lisbon, in A. Bausch and B. Schwenker (eds.), *Handbook Utility Management* (Berlin, Heidelberg: Springer, 2009).

6.2.1 Enforcement of EU Competition Law in the Energy Sector

Prior to the Lisbon Treaty, the Commission exerted its influence in the energy sector using its competences in the areas of single market and environment. EU competition law has been the instrument to bring energy agreements under the increased scrutiny of the Commission.[14] Competition law instruments are used by the Commission and Member States' competition authorities to deal with competition cases and enforce EU competition law, such as antitrust regulation, merger control, and State aid regulation.

Following the 2007 Energy Sector Inquiry,[15] the Commission's focus on competition law enforcement continued with many cases brought under allegations of anticompetitive use of long-term natural gas contracts and corresponding capacity reservations. For instance, in 2009 the Commission fined GDF Suez and E.ON €553 million for operating a market-sharing agreement which was concluded in 1975. Under this agreement, the parties agreed not to supply natural gas in each other's territories (French and German gas markets). This was a time where there were de facto market entry barriers in place.[16] However, the companies continued to implement the 1975 agreement even after the Second Energy Package entered into force. The Commission found that this behaviour was anti-competitive and a very serious infringement of competition law.[17] The outcome of the case was immediate release of short- and long-term capacity. The decision made it clear that

---

[14]For a detailed examination of implementation of EU Competition law and policy in the energy sector, see Peter D. Cameron and Michael Brothwood, *Competition in Energy Markets: Law and Regulation in the European Union* (Oxford University Press, 2002) (for pre-Second Energy Package period) and Peter D. Cameron, *Competition in Energy Markets: Law and Regulation in the European Union* (Oxford University Press, 2nd edition, 2007).

[15]DG Competition Report on Energy Sector Inquiry, 10.01.2007, SEC (2006) 1724.

[16]Summary of Commission Decision of 8 July 2009 relating to a proceeding under Article 81 of the EC Treaty (Case COMP/39.401—E.ON/GDF) (notified under document C (2009) 5355 final) (2009/C 248/05).

[17]In particular, the Commission found that the agreement between the two undertakings infringed Article 81(1) of the Treaty Establishing European Community (EC Treaty) (Article 101 of the TFEU), which prohibits concerted practices that restrict competition.

competition law imposes limits on the extent that dominant companies can reserve infrastructure capacity on a long-term basis. It has also shown that access issues remain among the most complicated, yet crucial, questions in the EU utilities sectors.[18]

On March 2009, the Commission also adopted a commitment decision against RWE AG, a natural monopoly, for concerns of abuse of dominant position in the German gas market. The commitment addressed possible foreclosure of RWE's competitors from access to its gas network (by way of a refusal to supply transportation capacity) and a possible margin squeeze to the detriment of RWE's competitors. The Commission offered a structural divestiture remedy to RWE, aimed at facilitating competition in this sector.[19] An efficient and non-discriminatory access to transport capacity is a pre-condition for the functioning of competition. Access can be required through sector specific regulation or on the basis of competition law. These cases coincided with the period of negotiations of the Third Energy Package, which establishes the basic rules and principles of third-party access.

However, even after the entry into force of sector specific regulation (Third Energy Package), the Commission's activities in enforcement of competition law in the energy sector continues, although mainly in relation to State aid rules under Article 107 TFEU. For instance, in 2012 the Commission initiated formal proceedings against Gazprom to investigate whether the company divided gas markets by hindering the free flow of gas across Member States, prevented the diversification of supply of gas, and whether it imposed unfair sales prices by linking the price of gas to oil prices in Central and Eastern European gas markets in breach of EU antitrust rules.[20] In May 2018, the Commission adopted

---

[18]Directorate General for Internal Policies, Policy Department A: Economic and Scientific Policy: Competition Policy and an Internal Energy Market, July 2017, available at: http://www.europarl.europa.eu/RegData/etudes/STUD/2017/607327/IPOL_STU(2017)607327_EN.pdf.

[19]The case was brought under Article 82 of the EC Treaty (Article 102 of the TFEU) which concerns abuse of dominant position. Commission Decision of 18 III 2009 relating to a proceeding under Article 82 of the EC Treaty and Article 54 of the EEA Agreement (Case COMP/39.402—RWE Gas Foreclosure).

[20]European Commission, Press release, Antitrust: Commission Opens Proceedings Against Gazprom, 04.09.2018, available at: http://europa.eu/rapid/press-release_IP-12-937_en.htm?locale=en.

its decision (four years after the case formally opened) and imposed a set of obligations on Gazprom. These obligations are, inter alia, aimed at enabling the free flow of gas without territorial restrictions in supply agreements and ensure that gas prices remain at a competitive level, for example by giving customers the right to demand a price adjustment from Gazprom if the price is higher than competitive benchmarks as they can be found in western European gas hubs (spot prices).[21]

6.3   Internal Market Legislation

The internal energy market, as the regulatory and infrastructure framework, should allow the free flow and borderless trade of gas and electricity across the EU. The Third Energy Package aimed to achieve the internal market by 2014. As revealed in Chapter 4 the EU budget also provided €3.7 billion of financing for energy infrastructure between 2007 and 2013, with a further €5.3 billion covering the period between 2014 and 2020 under CEF Funding, which is renewed until 2027 with a new energy budget of €8.7 billion.[22]

Alongside efforts to establish internal regional economic integration and a co-ordinated European approach *vis-à-vis* countries outside the EU, the Commission has simultaneously pursued a policy for construction of a regional energy market and common energy policy area within the EU. This was the original intention of the drafters of the European Coal and Steel Community.

National energy markets in almost all Member States were originally controlled and dominated by state monopolies since, given the EU's size and social components, controlling a vertically integrated single entity that had the right to produce, transport and supply natural gas (and also electricity) to end users was perceived as an easier option.

---

[21]CASE AT.39816—Upstream Gas Supplies in Central and Eastern Europe, Antitrust Procedure, Council Regulation (EC) 1/2003, Article 9 Regulation (EC) 1/2003, 24.05.2018.

[22]Special Report No. 16/2015: Improving the Security of Energy Supply by Developing the Internal Energy Market: More Efforts Needed, available at: http://www.eca.europa.eu/Lists/ECADocuments/SR15_16/SR_ENERGY_SECURITY-EN.pdf.

Energy liberalisation in the US, which took place in the 1970s and a decade after in the UK, inspired several EU level Directives and Regulations as well as Guidelines enacted from the late 1990s, in order to introduce energy market liberalisation in the EU. This legislation, inter alia, covered electricity and gas markets and the renewables sector. The idea was to move towards natural gas becoming a commodity rather than being merely a public service for society. Naturally, there was no immediate competition in the market, given the infrastructure inherited from the era of natural monopolies. An internal market, where natural gas (and electricity) would flow freely, requires that there is free capacity available within fixed-network bound capacity to allow entry by entities from other EU Member States and the technical standards to match.

The creation of an internal market was therefore a gradual process implemented in a step-by-step manner. The deficiencies were addressed along the way by the enactment of new legislation and policy objectives.

6.3.1 Transit Directive and the First Gas Directive

The first liberalisation legislation to pursue a new European energy policy and create a European energy market were the Electricity and Gas Directives of 1996 and 1998, respectively. The first Gas Directive co-existed with the Transit Directive which remained in force until 2004. This implies that until 2004, aside from transit through third countries outside Europe, the concept of 'transit' existed between the then 17 EU Member States, and natural gas activities were regulated at the EU level under the two directives. The transit directive applied only with respect to transit of natural gas involved in trade undertaken by a list of major companies annexed thereto. The definition of transit explicitly referred to high-pressure grids and applied in instances where transport involved crossing of only one intra-border.[23]

---

[23]Art. 2(1) Transit Directive.

The first Gas Directive did not apply to 'transit' but only to transmission (including transportation) inside a member State (without an offtake) through a high-pressure pipeline network used to deliver natural gas produced in a country of origin to the customers.[24] However, it set forth an access regime under which the then Member States had the option to allow negotiations or to regulate access based on published tariffs, terms and obligations.[25] TSOs could only refuse access to the system on the basis of the lack of capacity or where they would cause serious economic and financial difficulties due to foregoing contractual obligations or where such access would prevent them from carrying out their public-service obligations.[26] Additionally, Member States could deny access in the event of 'sudden crises'.[27] This Directive was to be implemented within two years; however, ultimately it was implemented inconsistently, in an environment where Member States continued to enjoy freedom in shaping national energy policies.

This period corresponds with the development of the EU's Energy Charter initiative, driven by the aspiration to promote cooperation with the energy rich transition economies of the former Soviet Union (FSU) states,[28] with a view to securing future supplies from the vast natural gas reserves of the Caspian and the Caucasus regions. The roots of the Energy Charter date back to a political initiative launched in Europe in the early 1990s, at a time when the end of the Cold War offered an unprecedented opportunity to overcome previous economic divisions. Nowhere were the prospects for mutually beneficial cooperation clearer than in the energy sector, and there was a recognised need to ensure that a commonly accepted foundation be established for developing energy cooperation among the states of Eurasia. Based on these considerations, the Energy Charter process was born. Notably, the most significant non-EU transit countries, Russia, Belarus and Ukraine were

---

[24]Art. 2(3) First Gas Directive.

[25]Art. 15 and Art. 16 First Gas Directive.

[26]Art. 17 First Gas Directive.

[27]Art. 24 First Gas Directive.

[28]Armenia, Azerbaijan, Belarus, Estonia, Georgia, Kazakhstan, Latvia, Lithuania, Moldova, Russia, Ukraine, Uzbekistan.

all included in the Energy Charter constituency. In addition, The Trans-Mediterranean pipeline from Algeria through Tunisia to Sicily (Italy) started flowing gas to mainland Italy in 1983. Also, the GME pipeline from Algeria through Morocco to Spain and Portugal was completed at the end of 1996.

The Energy Charter Treaty (ECT) entered into force in 1998.[29] The key purpose was not to reduce the risk of supply interruptions, supply scarcity or security of energy infrastructure. This is due to the fact that, save the two World Wars and the 2006–2009 crises, interruption occasions were infrequent. Rather, the initiative was focused on the establishment of an energy market compromising those Eurasian states linked by energy flows between the east and west (i.e. all EU Member states, Visegrad countries, the CIS, Russia, Turkey, Ukraine, Turkmenistan, Kazakhstan, Georgia and Azerbaijan). The Energy Charter aimed to create a level playing field for energy trade facilitation and market integration, the development of upstream and mid-stream activities and to establish a forum within which the EU could pursue enlargement of its energy policies towards the east while the rest of the constituency could voice their justified concerns over exploration, exploitation and transport of their natural resources. Given the importance of the question of transit the ECT was seen as the best available instrument to address these issues. The endeavour to agree on a separate 'multilateral transit framework' in the form of a special additional agreement—a Transit Protocol—to be applied within the ECT constituency began in 2000. However, the negotiation process was suspended in 2003.

At the same time, policy perception shifted in the EU resulting in a change of policy which aimed to integrate energy markets and break the monopoly of state-owned natural gas utilities, including vertically integrated incumbents, particularly those owning and operating natural gas pipelines. This regulatory process involved unbundling their assets along the gas value chain. In other words, separating transmission and

---

[29]ECT was opened for signature in Lisbon 17 December 1994 and entered into force on 16 April 1998. See ECT's home page available at: http://www.encharter.org/.

distribution networks from sales and supply to provide access on a fair and transparent basis, increasing competition and enhancing market liquidity.

6.3.2 The Second Energy Package

The Second Energy Package, adopted in 2003 consisted of a Gas Directive (the second Gas Directive)[30] and Regulation 1775/55/EC,[31] whilst repealing both the first Gas Directive of 1991 and the Transit Directive. Through this legislation the EU Commission hoped to facilitate market integration and pursue a strategy to remove the concept of transit inside EU borders. The key term 'transmission' is defined in Article 2.3 of the second Gas Directive as:

> *'Transmission' means the transport of natural gas through a high-pressure pipeline network other than an upstream pipeline network with a view to its delivery to customers, but not including supply.*

A more detailed definition was provided in Article 2 of the Gas Regulation (1775/55/EC):

> *'Transmission' means the transport of natural gas through a network, which mainly contains high pressure pipelines, other than an upstream pipeline network and other than the part of high pressure pipelines primarily used in the context of local distribution of natural gas, with a view to its delivery to customers, but not including supply.*

The second Gas Regulation prescribed a move towards further liberalisation and aimed to increase competition within the European energy network. The availability of transport capacity is a key requirement for the development of a competitive internal market in natural gas. The

---

[30]Thomas W. Walde (ed.), *The Energy Charter Treaty: An East-West Gateway for Investment & Trade* (1996).

[31]Regulation (EC) No. 1775/2005 of the European Parliament and of the Council of 28 September 2005 on conditions for access to the natural gas transmission networks.

access regime of the second Gas Directive excluded the possibility of negotiated access to transmission networks and imposed regulatory, third-party access based on non-discriminatory and cost-reflective published tariffs (mandatory third party access) as the only option.[32] As such, member States were called to designate 'National Regulatory Authorities' (NRA) that, inter alia, have the responsibility to fix or approve the tariffs or the methodology underlying their calculation.[33] The conditions for denial of access largely remained the same as those listed in the first Gas Directive.[34] Notably, the second Gas Directive introduced an exemption mechanism for major new infrastructure and for significant increases of existing capacity which would entail immunity from the application of mandatory third-party access and rules concerning tariffs, provided that stipulated conditions are met.[35]

This legislation brought stringent regulation, particularly relating to mid-stream and downstream natural gas activities, something which was not wholeheartedly welcomed by external producer states and domestic/foreign private companies active in the European gas markets.[36]

6.3.3 Exemption

Fixed and network bound energy infrastructure, in particular natural gas pipelines, are subject to economies of scale: the bigger the pipe, the sooner investments costs will be recovered. It is therefore not practical to build two pipelines along the same corridor instead of one. This makes pipelines an essential facility, one in the sphere of EU competition law, which means that in order to allow new market entry, third-party access is inevitable.[37] Third-party access is only granted

---

[32]Art. 18 Second Gas Directive.

[33]Recital 13 and Art. 25 Second Gas Directive.

[34]Art. 21 Second Gas Directive.

[35]Art. 22 Second Gas Directive.

[36]Sergei Vinogradov and Gokce Mete, Cross-Border Oil and Gas Pipelines in International Law, *German Yearbook of International Law* (2014), p. 56.

[37]Kim Talus, Just What Is the Scope of the Essential Facilities Doctrine in the Energy Sector? Third Party Access-Friendly Interpretation in the EU v. Contractual Freedom in the US, *Common Market Law Review*, Volume 48, Issue 5 (2011), pp. 1571–1597.

or refused on a transparent and non-discriminatory basis. On certain limited grounds third-party access can also be exempted (for a limited amount of time) via an exemption from the rule. This is provided in Article 22 of the Second Gas Directive. The exemption decision should also be taken transparently and without discrimination. The conditions to satisfy an exemption are as follows: (a) the investment must enhance competition in gas supply and enhance security of supply; (b) the level of risk attached to the investment is such that the investment would not take place unless an exemption was granted; (c) the infrastructure must be owned by a natural or legal person which is separate at least in terms of its legal form from the system operators in whose systems that infrastructure will be built; (d) charges are levied on users of that infrastructure; and (e) the exemption is not detrimental to competition or the effective functioning of the internal gas market, or the efficient functioning of the regulated system to which the infrastructure is connected.

However, national rules adopted to implement the Second Gas Directive differ from one Member State to another. Furthermore, as explained above, the exemption decisions were taken by relevant NRAs, while the Commission had a right to veto the decision. On the one hand, this might result in a discriminatory decision, one that may not allow for an exemption despite the conditions being met. This would affect the economic viability of a given project. On the other hand, a Member State might be too generous in giving exemptions, thus hampering competition and market integration. It is important to note that the EU institutions enjoy certain discretionary powers in deciding which projects pose danger to the security of supply and which do not.[38]

---

[38]Luca Franza, Outlook for Russian Pipeline Gas Imports into the EU to 2025, Clingendael International Energy Programme, The Hague, the Netherlands (2016).

6.3.4 Third Energy Package

The last phase[39] of the establishment of a competitive and single regulatory European energy market was concluded in 2009 with the entry into force of the Third Energy Package. The Third Energy Package included a Gas Directive (alongside a corresponding Directive and Regulation on Electricity) with common rules for the internal market in natural gas,[40] the Gas Regulation with conditions for access to the natural gas transmission networks,[41] and Regulation on the establishment of ACER.[42] The Third Energy Package emerged following the 'Internal Market Review'[43] and the 'Sector Inquiry Report'[44] reflecting the Commission's dissatisfaction with how the previous set of energy legislation had been implemented. These reports found that the unbundling requirements imposed on system operators were largely inadequate and that the market was prone to vertical foreclosure and thus lacked the required integration and transparency.

The overall innovations that came with the Third Energy Package include the following:

- ownership unbundling;
- Third Country Clause—that requires non-union undertaking to unbundle;
- further independence of NRAs;
- increased and institutionalised cooperation among the NRAs via ACER;

---

[39]Although the Clean Energy Package followed the Third Energy Package, it is not a marker regulation per se. The first three packages can be described as market packages because the aim is to create a market; with the Clean Energy Package the objective is different as it focus more towards consumers. Further, the electricity directive that came under the Clean Energy Package is not a new regulation but a revision of the rules.
[40]Gas Directive.
[41]Regulation 1775.
[42]Acer Regulation.
[43]COM (2007) 724 of 20 November 2007.
[44]MEMO/07/15 of 10 January 2007.

- increased and institutionalised cooperation and coordination among European TSOs via ENTSOs;
- consumer protection; and
- non-discriminatory access to LNG and storage

The rationale behind the EU's gradual move towards increased integration is explained in greater detail elsewhere.[45] During the period between the emergence of the Second Energy Package and Third Energy Package,[46] significant changes took place. First, both the commodities (price) markets and capacities markets in Europe changed. This was associated with changes in the level of actors involved in the energy value chain (from vertical integration, where producers would engage in trade with one company, to unbundling and a multitude of actors) particularly under Regulation 1775.[47] Second, the delivery points of natural gas for external suppliers changed from outside the EU to within it as a result of changed borders (enlargement of membership from 17 to 28 Members between 2004 and 2013) bringing them closer to Russia (the largest external supplier). At the same time, although most significant technical and practicable issues were agreed upon by 2009, efforts to agree on all encompassing rules governing transit as an additional protocol to the ECT, which could have been based on a compromise which reflected concerns of producers, transit and consumer states, stalled.

These developments brought the core EU energy markets and third countries with energy infrastructure links to the EU under the rule of the Third Energy Package, which set a higher threshold of regulation and market intervention than the ECT, thus creating a gap between them. This higher regulatory threshold came, as did the Second energy package, with an option to soften the rules under various available derogations to enable the development of strategic and critical major natural gas infrastructure and for significant increases in existing capacity. As explained in Chapter 5 on infrastructure, currently, most natural gas pipelines are developed and operated under diversions from the rules

---

[45]Kim Talus, *EU Energy Law and Policy: A Critical Account.*
[46]The Third Energy Package Includes: Acer Regulation, Regulation 1775, Gas Directive.
[47]Regulation 1775.

providing exemptions from the application of mandatory third-party access, unbundling and provisions concerning tariffs, provided that stipulated conditions are met.[48] At the same time regulations acknowledged the long-term structure of existing natural gas supplies, as it was not possible to retrospectively challenge the contracts signed with external suppliers (Algeria, Russia, and also Norway) and that security of supply could not otherwise be sustained (existing contracts later challenged by the 2019 Gas Directive amendment, further discussed below).

The question thus arises: are the rules serving their intended purpose? In other words, do they contribute to the security of supplies, increase market competition, liquidity and benefit end users (consumers) by decreasing prices and enhance market integration while at the same time meeting other EU energy and climate policy objectives? These objectives include lowering carbon emissions by substituting other fuels with natural gas in power generation, industry and households and ensure sustainability.[49]

Serious concerns over the credibility and predictability of EU energy policy and regulation reached a new peak following the third and most recent Russia–Ukraine transit crisis, on top of the ongoing disagreement between Russia and the EU over the transition of EU energy markets under the Third Energy Package. The most significant reason for Russian opposition to this regulation lies in its so called 'Third Country Clause', a provision that obliges external producers of natural gas to abide by internal market rules.

6.4 Third-Party Access

Since third-party access is mandatory and based on non-discriminatory published tariffs,[50] the access regime, which includes circumstances that tolerate denial of access, remained more or less the same. According to

---

[48]Art. 36 Third Gas Directive.
[49]EC Press release, 2030 Climate and energy Goals for a Competitive, Secure and Low-Carbon EU Economy, availabe at: http://europa.eu/rapid/press-release_IP-14-54_en.htm.
[50]Art. 32–35 Gas Directive.

Article 13 of Regulation 715, Member States may elect to apply market-based tariffs or empower NRAs to approve tariffs proposed by relevant TSOs. The fundamental principles applicable to access conditions were introduced in Regulation 1775 and retained under Regulation 715. However, rules (as opposed to principles) concerning CAM and congestion management, are to be devised under Network Codes, developed by ENTSOG, in line with 'framework guidelines' prepared by ACER. The first CAM Network Code was published on 14 October 2013, to be applicable from 1 November 2015 (now repealed and replaced by an amendment, CAM NC).[51] It covers only interconnection points, and the CAM is based on auctions. On 26 March 2014, the Commission adopted the second EU-wide gas Network Code on Gas Balancing of Transmission Networks applied from 1 October 2015.[52]

6.4.1 Unbundling

The second Gas Directive obliged Member States to designate TSOs,[53] and required the certification of designated TSOs by the NRA of the Member State where the transmission system in question is located.[54]

Certification was conditional upon satisfying one of the three unbundling models: ownership unbundling (OU), independent system operator (ISO) or independent transmission operator (ITO).[55] Therefore, another important amendment to the former Gas Directive came into effect. In simplified terms, OU means that a vertically integrated production or supply undertaking may not own or retain decisive control

---

[51]Commission Regulation No. 984/2013 of 14 October 2013 establishing a Network Code on Capacity Allocation Mechanisms in Gas Transmission Systems and supplementing Regulation (EC) No. 715/2009 of the European Parliament and of the Council [2013] OJ L273/5.

[52]Commission Regulation (EU) No. 312/2014 of 26 March 2014 establishing a Network Code on Gas Balancing of Transmission Networks [2013] OJ L91/15.

[53]Art. 7 Second Gas Directive.

[54]Art. 10 Gas Directive.

[55]Art. 9 Gas Directive.

over the transmission operations. Under the ISO system, the owner of supply or production activities upstream may not, at the same time, operate the transmission system. Although the upstream undertakings may own the transmission infrastructure assets, the system operator itself must be a separate entity fully independent from such producers or suppliers. The ITO system permits the transmission operator to be the same entity as the producer/supplier undertaking and retaining ownership. However, more stringent regulations are enforced in order to ensure that such an ITO is managed independently, supervised adequately and that sufficiently deterrent fees are charged for any noncompliance. Member States have the right to determine the type of OU system applicable in its internal market.

A core competition rule, unbundling in principle aims to avoid conflicts of interest along the entire value chain, in particular between production, transport and supply chains, and is a necessary component of third-party access. Nevertheless, the so called third country clause caused some concern for multinational corporations and non-union undertakings, as well as the international finance community, due to the initial uncertainty it introduced. The geographical scope of the rules was not entirely clear (and arguably became clearer but more complex under the 2019 Gas Directive Amendment).

Unbundling also allows a non-EU undertaking to operate directly inside the retail market. For instance, Gazprom Marketing and Trading has been able to supply gas to industry and commerce since in 2006.[56] This company is a 100% subsidiary of Gazprom Export, the export arm of OAO Gazprom, the world's largest gas producer. It is worth noting that the Gas Directive maintained the possibility for exemptions, now also applicable to the unbundling obligation.[57] Investors from third countries that satisfy the conditions listed may also apply for an exemption.

---

[56]See Gazprom MT's website available at: http://www.gazprom-mt.com/.
[57]Art. 36 Gas Directive.

6.4.2 Transmission System Operator (TSO)

The operation, maintenance, modernisation and expansion of a given natural gas transmission system is performed by TSO(s). Traditionally, along the Eurasian energy corridor, stretching from Europe towards Central Asia, TSOs were public entities, often vertically integrated in the national energy company (monopoly) that was also active in the upstream or wholesale part of the sector. They would operate the system as a common carrier but in most cases subject to supervision by a ministerial, competition or regulatory authority.

Privatisation of TSOs and unbundling came with the introduction of competition and liberalisation in the EU. This opened the door for the possibility of TSOs to be international, including ownership by non-EU companies, although they still had to have the public service obligations component. There has been significant progress in the privatisation process in the EU, however a number of challenges may still arise from a competition law perspective. For instance, 66% of the shares of the Greek national TSO, DESFA, was acquired by Azerbaijan's state-run company, SOCAR, via a tender in 2013.[58] The process was stalled by an investigation launched by the EC in 2014 in accordance with EU Merger Regulations. This is because DESFA was not only the TSO but also the operator of the sole LNG Terminal in Greece, and the Commission considered that market foreclosure could become an issue as SOCAR has its own production and participates in downstream and upstream markets in Greece. Hence, an opportunity to privatise the Greek national market stalled due to the stringent application of competition rules.

If the Commission had given green light to this acquisition, SOCAR's certification as an ITO had to be carried out in accordance with Article 11 of the 2009 Gas Directive (which extends requirements

---

[58]European Commission Opinion of 17.10.2014 correcting Opinion C(2014) 5483 final of 28 July 2014 pursuant to Article 3(1) of Regulation (EC) No. 715/2009 and Article 10(6) and 11(6) of Directive 2009/73/EC—Greece—Certification of DESFA, Brussels, 17.10.2014, C(2014) 7734 final.

contained in Article 10 to non-EU undertakings). This also shows that the rules are applicable to all third country undertakings and not only Gazprom (i.e. South Stream). It is interesting to note that the Russian Federation claimed that the different criteria imposed by the Third Energy Package on TSOs not owned by EU Member States is discriminatory and breaches the principles of national treatment, fair and equitable treatment and most-favoured nation treatment of the ECT and Word Trade Organization (WTO).[59] In fact, Russia requested the establishment of a panel under the WTO on 28 May 2015 (see more below).[60]

Following the collapse of the previous attempt to privatise DESFA in 2016, privatisation was postponed and the acquisition was only completed in December 2018 when a consortium consisting of Snam (60%), Enagás (20%) and Fluxys (20%) took a 66% stake in DESFA from the Hellenic Republic Asset Development Fund (HRADF) and Hellenic Petroleum for a consideration equal to €535 million.[61]

Outside the regulatory area of the Third Energy Package, as discussed in Chapter 3, cross-border natural gas and oil pipelines are commonly built as merchant pipelines where the transit/transport systems are built by private capital, such as the Baku-Supsa or Baku Ceyhan pipelines. In these cases, the entire, or a significant percentage of the capacity, may be reserved for one or a limited number of shippers (uncommon carrier system), as the primary objective would be to recover the upfront investments. Competition concerns may also arise instead due to ex-ante regulation, such as Articles 10 and 11 of the 2009 Gas Directive; this will be ex-post control.

---

[59]WT/DS476/2.a, European Union and Its Member States—Certain Measures Relating to the Energy Sector, available at: https://www.wto.org/english/tratop_e/dispu_e/cases_e/ds476_e.htm.

[60]WT/DS476/2.a, European Union and Its Member States—Certain Measures Relating to the Energy Sector, available at: https://www.wto.org/english/tratop_e/dispu_e/cases_e/ds476_e.htm.

[61]Press release, Fluxys, The Snam, Enagás, Fluxys Consortium Completes the Acquisition of 66% of the Greek Operator DESFA, 20.12.2018, available at: https://www.fluxys.com/en/press-releases/fluxys-group/2018/181220_press_acquisition_desfa.

6.4.3 Third Country Clause

Perhaps the most significant innovation of the Third Energy Package is the so-called third country clause. As explained above, this obliges persons from a third country or countries acquiring control of the transmission system or the TSOs to satisfy the certification conditions, in other words, meet the terms of the unbundling provisions.[62] An additional condition, applied in respect of third countries, is that investment by third country entities should not endanger the security of supply of a Member State in question or of the EU. In such a case, the Member State has the ultimate right to refuse requests for certification.[63] In addition to the unbundling obligation, potential TSOs from third countries, once certified, has to abide by the mandatory third-party access and tariffs principles, as well as the Network Codes of the EU energy *acquis*.

The rationale of the third country clause was that if third countries who are active in both production, transport and trade of energy were not subject to the same rules as the EU undertakings, this would weaken the competition benefits aimed at breaking vertical integration. Otherwise, the international giants would go on and purchase unbundled EU undertakings (as the case of DESFA indicates—both SOCAR and Gazprom bid for its acquisition). Another reason is that Article 11 of the 2009 Gas Directive is expected to promote the opening of markets particularly within the immediate neighbourhood (i.e. Russia and Turkey).

6.4.4 Exemptions and Derogations

It is important to note that the Gas Directive maintained the exemption possibility, now additionally covering the unbundling obligation.[64] Investors from third countries that satisfy the conditions listed may also apply for an exemption. Fifteen cases of exemptions for new gas pipeline infrastructures have been decided as of June 2018, four of which

---

[62]Art. 11(5)a Gas Directive.
[63]Art. 11(5)b Gas Directive.
[64]Art. 36 Gas Directive.

were exemption requests from the Third Energy Package.[65] In the subsequent sections, three of these projects where an exemption is relevant are presented. These case examples have been mentioned throughout the previous chapters of this book but this section provides a detailed account of the exemption decision making procedure.

6.4.4.1   North Stream

In the case of the North Stream pipeline, Gazprom had applied for an exemption for the pipelines onshore extension in Germany, the OPAL and NEL (Northern European natural gas pipeline) sections. The German NRA BNetzA granted a 100% third party access exemption for the capacity of the pipeline for 22 years, which meant that the full capacity was dedicated to Gazprom. The exemption decision was accompanied by various conditions imposed on Gazprom to prevent capacity hoarding and ensure proper congestion management. However, the Commission limited the exemption by half in 2009 as it considered the project not beneficial to competition.[66]

After years of negotiation, the Commission and Gazprom agreed that the full capacity could only be utilised by Gazprom if there was no demand by a third party. This decision of the Commission was expected to be announced in 2014, however as a result of the political tensions related to the dispute over Crimea and ongoing conflict in eastern Ukraine, it has never been sealed. The Commission cited technical issues as a reason for the delay.[67] In 2016 the German NRA notified

---

[65]The list of exemption decision from the Third Energy Package could is accessible via: https://ec.europa.eu/energy/sites/ener/files/documents/exemption_decisions2018.pdf.

[66]Tjarda van der Vijver, Third Party Access Exemption Policy in the EU Gas and Electricity Sectors: Finding the Right Balance Between Competition and Investments, in Martha M. Roggenkamp et al. (ed.), *Energy Networks and the Law: Innovative Solutions in Changing Markets* (Oxford University Press, 2012).

[67]European Commission Delays Ruling on New Gazprom Bid for German Gas Link Opal Use, Platts, 14 July 2016, https://www.platts.com/latest-news/natural-gas/brussels/european-commission-delays-ruling-on-new-gazprom-26492211.

the Commission and requested changes to the 2009 exemption decision due to market changes in the 7 years since the pipeline was constructed. The German NRA requested removing the 50% capacity limitation and for it to be made available for firm capacity booking by all companies including the dominant players. In its decision in October 2016, the Commission decided that the short-term available gas capacity at the Gaspool hub ('FZK capacity') would increase by 20% and allowed undertakings or groups of undertakings with a dominant position in the Czech Republic or which control more than 50% of natural gas arriving at Greifswald, including Gazprom, to be able to bid for this capacity at the base price at auctions organised by the PRISMA platform. Following this, BNetzA, OPAL Gaztransport, Gazprom and Gazprom Export signed a settlement agreement in November 2016 to implement the Commission's exemption decision (most capacity booked in PRISMA is still long-term but short-term capacity is increasing).

In early 2017, capacity sharing of the OPAL pipeline was still unresolved and the issue became more complicated than it was in December 2016. The Polish state-run gas company PGNiG sued the European Commission over the decision, claiming that the November 2016 settlement agreement decision would have a negative impact on the security and competitiveness of gas supplies to Poland.

The case was referred to the European Court of Justice (ECJ)[68] and upon PGNiG's request to the ECJ, it was granted a provisional suspension in December 2016 from execution of the October 2016 exemption decision. However, the ECJ rejected the applications for suspension later in July 2017 as the existence of

---

[68]PGNiG's German subsidiary (PGNiG Supply & Trading GmbH) vs EC (4 December 2016, case T-849/16); Poland vs EC (16 December 2016, case T-883/16); PGNiG vs EC (1 March 2017, case T-130/17).

two contracts concluded by Gazprom for supply and transit contract for the transport of natural gas via *the Polish section of the Yamal-Europe pipeline to supply the Western European markets (including Poland) gas until the end of 2022 and until 2020 respectively guaranteed Gazprom's deliveries to the Polish market until the aforementioned dates.*[69]

Meanwhile auctions of the OPAL capacity continued. However, the Court delivered its decision on 10 September 2019.[70] The General Court annulled the Commission decision approving the modification of the exemption regime for the operation of the OPAL gas pipeline, noting that the Commission's decision was adopted in breach of the principle of energy solidarity (the principle of solidarity requires Member States to take into account the interest of other Member States and of the EU with respect to their decisions in their energy market which may have a cross-border impact). This means that Gazprom still reserves the right to use 50% capacity of OPAL pipeline but it will no longer be able to participate in auctions for the remaining 40%. While it is unlikely, the European Commission has two months to decide whether or not to appeal against the decision.

The 2016 suspension decision of the ECJ has been perceived by think-thanks as proof of politicisation of the EU's gas legislative and regulatory framework, as it has shown that the initial 50% cap on Gazprom's capacity in 2009 was not based on competition law or the *acquis*. The latest attempt by Poland to challenge this decision is also criticised as another attempt to politicise the EU's rules-based regulatory decision-making.[71] However, a major reason for this rather chaotic and complex situation is the lack of upfront regulations on new capacity development (at the time

---

[69]The President of the General Court rejects the applications for a stay of execution of the Commission's decision that 50% of the transport capacities of the OPAL gas pipeline are to be subject to a bidding procedure, General Court of the European Union, Press release No. 83/17 Luxembourg, 21 July 2017, Order of the President of the General Court in Cases T-849/16 R, T-883/16 R and T-130/17 R PGNiG Supply & Trading GmbH, Poland, and Polskie Górnictwo Naftowe i Gazownictwo S.A. v Commission.

[70]Press Release, General Court of the European Union no 107/19, Luxembourg, 10 September 2019, Judgment in Case T-883/16, Poland v Commission, available at: https://curia.europa.eu/jcms/upload/docs/application/pdf/2019-09/cp190107en.pdf.

[71]Katja Yafimava, *The OPAL Exemption Decision: Past, Present, and Future* (OIES, 2017).

of construction of the North Stream pipeline in 2009 till now, August 2019) and the existing system's gaps allowing for arbitrary decisions. This is because the criteria for exemption is defined broadly and does not require a qualitative assessment, giving the Commission and the NRA a degree of discretion. Furthermore, the exemptions are not automatic and could be challenged by the Commission and concerned NRAs. The exemption mechanism under Article 36 of the 2009 Gas Directive was meant to be applied scarcely. The case examples since its entry into force however indicate that the exemption route has become almost the default rule under which a cross-border natural gas pipeline could be constructed.

The OPAL pipeline is not the only example of arbitrary decision making in implementation of the exemption route. The Nabucco project, for instance, received an exemption from the application of the then Second Energy Package despite the absence of potential supplies from third countries to fill the pipeline and a lack of commerciality (and funds to finance the project). The exemption was granted as the project was politically supported by the Commission. It could also be argued that the TAP pipeline should not have been granted an exemption from the Third Energy Package as it already benefited from its PCI status with accelerated permitting procedures and financial support. The TAP project is also supported by the Commission politically as part of its supply diversification policy.

The entry into force of the CAM Network Code in April 2017 (discussed further below) should reduce new pipeline projects' needs to resort to the exemption regime. That being said, the proposed Gas Directive Amendment discussed in this Chapter introduces a new set of exemptions and derogations to the Gas Directive, which is at odds with the EU's well-established policy of phasing out the exemptions regime.

6.4.4.2 TAP: A Project of Common Interest

TAP is an interconnector linking with the TANAP pipeline bringing Caspian gas resources to Europe as part of the Southern Gas Corridor. The specifics of this mega project was introduced in Chapters 4 and 5 of this book, namely that the project has already been selected as a PCI: hence, it benefits from the accelerated granting of necessary licences and permits. The TAP pipeline also received an exemption from certain provisions of the Third Energy Package, including a third-party access exemption for 25 years for the initial capacity of 10 bcm/y (i.e. a maximum of 50% of the total capacity of the project) for supplies from Azerbaijan under the relevant Shah Deniz gas sales agreements.[72] The exemption decision has subsequently been prolonged as the start date of the project has been "*delay*[ed] *beyond control of the person to whom the exemption has been granted*".[73]

The TAP exemption is important for the Turk Stream pipeline project as both of these projects are aimed to supply the same market in Southern Europe. The first line of Turk Stream will supply Turkey with Russian gas from 2020, with 15.75 bcm replacing Russian gas volumes that are currently flowing via Ukraine, Romania and Bulgaria via the Western Route. However, its second parallel line of the same volume is planned to stretch to Southern Europe when there is a demand (now foreseen to land on the border of Bulgaria instead of Greece). This second line could potentially connect to the TAP pipeline and utilise the 10 bcm idle capacity via a capacity expansion (although the EU's political will to give a green light to the second line remains rather dubious).

The legal framework of the TAP consists of separate bilateral, private law, host country agreements between Greece and TAP and between Albania and TAP. These agreements are supported by some additional agreements, such as the pricing agreements between TAP's home country Switzerland and the host governments to avoid double taxation on revenues. The TAP pipeline is important for gasification of the Balkan region and for replacing volumes in South Europe currently supplied through Ukraine from Russia as it is Russia's intention to terminate its transit and supply contracts with Ukraine from 2019.

---

[72]EC Decision of 16.5.2013 on the exemption of the Trans Adriatic Pipeline from the requirements on third party access, tariff regulation and ownership unbundling laid down in Articles 9, 32, 41(6), 41(8) and 41(10) of Directive 2009/73/EC, https://ec.europa.eu/energy/sites/ener/files/documents/2013_tap_decision_en.pdf.

[73]EC Decision of 17.3.2015 prolonging the exemption of the Trans Adriatic Pipeline from certain requirements on third party access, tariff regulation and ownership unbundling laid down in Articles 9, 32, 41(6), (8) and (10) of Directive 2009/73/EC, https://ec.europa.eu/energy/sites/ener/files/documents/2015_tap_prolongation_decision_en.pdf.

6.4.4.3 South Stream

As mentioned previously in this book, the South Stream pipeline is an aborted project. The process that led to its cancellation is an interesting one to single out and review as part of an investigation into the effectiveness of the legal and decision-making framework for gas infrastructure projects in the EU. As such it requires further discussion. The South Stream pipeline was to be constructed between Anapa, Krasnodarskiy Krai, on the Russian Black Sea shore and run 900 km through the Black Sea via the Turkish Exclusive Economic Zone (EEZ). This pipeline project envisaged the construction of four parallel lines with a total capacity of 63 bcm a year.

Gazprom did not apply for an exemption for the South Stream pipeline. Instead, Gazprom signed bilateral intergovernmental agreements with countries along the pipeline route: Russia, Bulgaria, Serbia, Hungary, Greece, Slovenia, Croatia and Austria. These were supplemented by agreements on cooperation between Gazprom and authorised national companies. Several joint companies based on essentially equal shares were established by Gazprom with local partners for every national on-land section. These companies were registered in and incorporated under the applicable legislation of each State. The initial 15.75 bcm capacity of the pipeline was scheduled to be completed in 2015. The agreements, however, were called for renegotiation[74] by the EU on the grounds that they did not comply with the unbundling and third-party access requirements of the 2009 Gas Directive and that the setting of tariffs was not in accordance with the said Directive.[75] Perhaps the mandatory third-party access problem could have been solved at the development stage of the project instead of the operational stage. For instance, it may have been possible to estimate the demand of third-party shippers via a 'coordinated open season' and this may have been added to Gazprom's capacity so that the pipeline could be constructed to allow the total capacity.

---

[74]EC Decision 2012/994 of 25 October 2012 OJ 2012 L, 13. Press release, available at: http://www.europeanvoice.com/art./2013/december/south-stream-must-be-renegotiated-commission/78982.aspx.

[75]The construction of the pipeline nevertheless continues. The Third Energy Package does not contain a provision requiring termination of the construction. However if the participating States refuse to renegotiate the terms of the bilateral IGAs, once the hydrocarbon transmission

It was also reported that according to the IGAs tariffs were to be determined by 'the Company', whereas in this regard, the 2009 Gas Directive (paragraph 32) empowers the "[N]*ational regulatory authorities to fix or approve tariffs, or the methodologies underlying the calculation of the tariffs, on the basis of a proposal by the transmission system operator or distribution system operator(s)*". Moreover, the IGAs are said to have provided priority to subcontractors from Russian and Bulgarian State companies, a practice that would contradict EU competition legislation.[76] From the Russian perspective the IGAs were regarded as valid under general international law, as it is perceived to have superiority over EU law.[77]

Russia also claimed that it should be immune from the application of the Third Energy Package based on a 1997 EU-Russia Partnership and Cooperation Agreement (PCA)[78] which requires the parties to use their "*best endeavours to avoid taking any measures or actions which render the conditions for the establishment and operation of each other's companies more restrictive than the situation existing on the day preceding the date of signature of the Agreement*". However, with respect to energy, the PCA has only one provision that refers to cooperation in the field of energy and this is framed within the principles of market economy and the European Energy Charter. Since Russia's withdrawal from the ECT, this relationship is based on non-legal commitments and dialogue on

---

starts problems may arise on booking capacity by third parties. At this point the EU might start infringement procedures. See reporting by ITAR-TASS, Interfax, RFE/RL Brussels correspondent Rikard Jowziak, and B92.net on 13 December 2013, available at: http://www.rferl.org/content/eu-renegotiate-south-stream/25191193.html.

[76]Press release, available at: http://eurodialogue.org/EU-countries-ask-for-help-to-escape-from-South-Stream-mess.

[77]Russian Prime Minister Dmitry Medvedev told that "*Legal acts of the EU are considered to be national laws for the countries of the EU* [...] *(while) intergovernmental agreements, signed by* [countries] *of the EU are acts of international law. On the whole, there is a rule of precedence of international law over national law. We only understand it this way*", News release, available at: http://www.naturalgaseurope.com/south-stream-eu-laws.

[78]Agreement on partnership and cooperation establishing a partnership between the European Communities and their Member States, of one part, and the Russian Federation, of the other part—Protocol 1 on the establishment of a coal and steel contact group—Protocol 2 on mutual administrative assistance for the correct application of customs legislation, OJ L327, 28/11/1997, Article 34.

improvement of the quality and security of energy supply and formulation of energy policy and modernisation of energy infrastructure.[79]

6.5   Gas Target Model

The third wave of liberalisation of the EU's internal energy market, like any major legislative changes with an impact on the private sector, constituted an initial concern for all gas stakeholders including, perhaps predominantly, third country suppliers. This is due to the fact that the Third Energy Package introduces vital changes to cross-border transportation of energy. The model on which these changes are structured was presented in 2014 under the 'Gas Target Model' (GTM) initiated by the Council of European Energy Regulators (CEER) and developed by the Commission and the EU Energy Regulators to lead to the creation of a single internal gas market in the EU guided by the Third Energy Package.

The key feature of the GTM is that it restructured the European gas transportation model from a 'Point-to-Point' System, where gas was transported generally at the border of the market, to an Entry–Exit System with capacity bookings at different interconnection points in each entry–exit zone. In a Point-to-Point system, natural gas flows through a predetermined route, and in contrast to the Entry–Exit system, it cannot exit at any exit point.

The Entry–Exist system, which has been in operation since September 2014, created transportation zones where capacity might be traded independent from the commodity contract. This means that capacity at each exit point is freely allocable and could be supplied by any entry point. On track for full implementation of the Entry–Exit system,

[79]Agreement on partnership and cooperation establishing a partnership between the European Communities and their Member States, of one part, and the Russian Federation, of the other part—Protocol 1 on the establishment of a coal and steel contact group—Protocol 2 on mutual administrative assistance for the correct application of customs legislation, OJ L327, 28/11/1997, Article 65.

deviations were allowed due to physical infrastructure and contractual (existence of long-term contracts) constraints.[80]

As a result, separate transit systems continue to exist within the EU. For instance, between Bulgaria and Romania cross-border pipelines operate without a third-party access requirement.[81] Bulgartransgaz EAD, the gas transmission system operator, does not provide virtual reverse flow capacity in Negru Voda, where the Bulgarian system connects with the Romanian system, and Sikirokastrou, where it connects with the Greek system.[82] The Yamal pipeline in the Polish gas transit system also used a separate Entry–Exit system which covers one large trunk pipeline but there was little flexibility as there were no virtual points.[83] However, physical and virtual reverse flows have been established on the Yamal pipeline in 2016.

In Entry–Exit systems, the shipper has to separately book an entry and exit capacity at the city gate and contract with the DSO to deliver to final consumer. Apart from the risk of contractual mismatch, this system increases congestion risks due to individual bookings and increases transaction costs. Furthermore, as cross-border tariffs are not prohibited as in the electricity sector, in short distances, gas could traverse through multiple borders resulting in a pancaking effect.

---

[80]Study on Entry–Exit Regimes in Gas, Part A: Implementation of Entry–Exit Systems, DNV KEMA. By order of the European Commission—DG ENERGY, 11.12.2013, available at: https://ec.europa.eu/energy/sites/ener/files/documents/201307-entry-exit-regimes-in-gas-parta.pdf.

[81]Study on the conditionalities stipulated in contracts for standard capacity products for firm capacity sold by gas TSOs, ACER/OP/ADMIN/13/2017/LOT 2/RFS 01, Final Study, 03.04.2019, available at: https://www.acer.europa.eu/Official_documents/Publications/AnnexestotheACERReportontheconditionalitiesstipula/Underlying%20consultant%20study.pdf.

[82]Case C-198/12, JUDGMENT OF THE COURT (Fifth Chamber), Failure of a Member State to fulfil obligations—Internal market in energy—Gas transmission—Regulation (EC) No. 715/2009—Articles 14(1) and 16(1) and (2)(b)—Obligation to guarantee maximum capacity—Virtual reverse flow gas capacity—Admissibility), 5 June 2014. ACTION under Article 258 TFEU for failure to fulfil obligations, brought on 26 April 2012.

[83]Study on Entry–Exit Regimes in Gas, Part A: Implementation of Entry–Exit Systems, DNV KEMA. By Order of the European Commission—DG ENERGY, 11.12.2013, available at: https://ec.europa.eu/energy/sites/ener/files/documents/201307-entry-exit-regimes-in-gas-parta.pdf.

On the positive side, the GTM makes short-term bookings easier, which is welcome as shippers are reluctant to invest in long-term bookings due to long-term market uncertainties and primary capacity bookings from shippers decreasing due to a range of complex interactions in a changing energy market. The Entry–Exit system also has the advantage of responding to price signals via the establishment of liquid trading hubs. The hub system offers flexibility but lacks the reliability that came with long-term take-or-pay contracts which had provided investment and supply security by the sharing of risks between buyers and sellers. Hub trading depends on sellers' competition as opposed to bilateral negotiations but may pose supply risks where the seller is not able to get gas flows to entry points.

6.5.1 Network Codes

As stated previously, the Third Energy Package also introduced a new system for the establishment of binding European-wide Network Codes for cross-border and market integration issues consistent with non-binding Framework Guidelines. Regulation 715 anticipates the Network Codes to cover twelve broad market issues. The Network Codes, which are technical rules turning regulatory policies into operational norms, follow the principles and objectives of the Gas Regulation and aim at improving access arrangements in the internal market. The Network Codes are one of the key points of European harmonisation and the development of an integrated energy market. They offer an opportunity to simplify and to facilitate access to the European gas market.

The European Commission drafts an 'annual priority list' of areas to be included in the development of network codes for gas. It does this with the input of ACER[84] and ENTSOG.[85] Once the annual priority

---

[84]ACER which was established by rule 713/2009, is the European organisation of energy regulators.

[85]ENTSOG (European Network of Transmission System Operators for Gas), the European association representing TSO's and gathering 42 members in 25 countries, was created by directive 715/2009 in 2009.

list is established, ACER develops 'framework guidelines' which set principles for developing specific network codes within six months. These framework guidelines are used by ENTSOG as a basis to prepare a network code within 12 months which is submitted back to ACER who then delivers its opinion within 3 months. If ACER deems that the code fulfils its framework guidelines and the EU's internal market objectives, it sends the code to the Commission recommending its adoption. A network code becomes mandatory for Member States after a process called comitology.[86] Under the comitology, before the reforms introduced by the Lisbon Treaty, the Commission is assisted by a committee of Member States' representatives and decisions are taken according to different procedures defined in EU law. In the case of the regulatory procedure, the Commission submits its proposal to the Council only when the committee disagrees. The European Parliament must be informed and gives an opinion to the Council. The Council finally acts by qualified majority on the proposal. A comitology committee is not allowed to change the substance of a proposal, and they need a two-thirds majority in order to reject it. If approved by the committee, the EU network codes become annexed to the relevant regulation. If the proposal is rejected, the Commission may resubmit or present a legislative proposal on the basis of the EU Treaty. At any time, the Commission has the opportunity to use a process called 'direct comitology' allowing it to replace ACER and ENTSOG and to directly undertake the writing of regulatory procedures (e.g. Congestion Management Procedures, CMP).

The Commission studies it and then sends it to the Gas Committee made up of specialists from national energy ministries for an opinion.

---

[86]Under comitology, before the reforms introduced by the Lisbon Treaty, the Commission is assisted by a committee of Member States representatives and decisions are taken according to different procedures defined in EU law. In the case of the regulatory procedure, the Commission submits its proposal to the Council only when the committee disagrees. The European Parliament must be informed and gives an opinion to the Council. The Council finally acts by qualified majority on the proposal. A comitology committee is not allowed to change the substance of a proposal, and they need a two-thirds majority in order to reject it. If approved by the committee, the EU network codes become annexed to the relevant regulation. If the proposal is rejected, the Commission may resubmit or present a legislative proposal on the basis of the EU Treaty.

Once the Gas Committee accepts the draft network code, it is adopted via the so-called comitology procedure with the approval of the Council of the European Union and the European Parliament. Then, TSO's have to support these changes and make the necessary investments. ACER will then monitor implementation of the codes and can eventually submit recommendations to the Commission, the European Parliament and the Council when it deems this unsatisfactory.

The drafting phases of both the Framework Guidelines and the Network Codes closely involve important stakeholders, in particular via direct consultation. Accordingly, each market player contributes to writing those network codes. Many meetings are organised by ENTSOG during the writing process of codes, allowing all stakeholders to participate to the writing, although it has been argued by some that the time given for consultations has often been too short.

6.5.2 Capacity Allocation for Existing Capacity

The CAM NC regulates the allocation of unsold capacity at interconnection points, stipulating that, irrespective of the duration of the supply contract, capacity can only be booked for the next 15 years in the EU.[87] The rationale is that otherwise shippers could book the capacity for so long that they would foreclose short-term markets for small competitors.

There is another issue with the CAM NC for incremental capacity: the obligation to reserve 10% for short-term bookings.[88] This obliges investors to build a pipe with 111% of the requested capacity, allowing shippers to bid for capacity that is not yet built. It may be argued that the risks for the investor therefore would be about 10% higher.

Existing capacity in non-exempted natural gas pipeline projects was, by default, allocated by auction under CAM NC. However, an

---

[87]CAM NC, Art. 11(3).
[88]CAM NC, Art. 8(8).

'Incremental Capacity' chapter, in the form of a CAM NC amendment, has been introduced and entered into force on 6 April 2017. The issue of incremental capacity is closely related to the different Network Codes discussed below.

Under the auction procedure of the CAM NC, a network user can book yearly, quarterly, monthly, daily and within-day standard capacity products. The TSOs are called to provide details on the available interconnection capacity[89] and shippers to submit bids and offers for cross-border capacity. Where the capacity contract underlying a supply contract expires before the supply contract expires, the shipper has to participate in the CAM auction process to secure new capacity. Capacity is explicitly allocated to shippers via one or a limited number of joint web-based booking platforms operated by TSOs or third parties.

Furthermore, the capacity at interconnection points under the CAM NC can be offered as 'bundled capacity' or 'unbundled capacity'. Bundled capacity means the same level of entry and exit capacity on a firm basis at both sides of an interconnection point.[90] Although the TSOs have a duty to maximise the quantity of bundled capacity offered,[91] capacity at each entry and exit point may not match in all cases.[92] only in such cases may unbundled capacity be offered. Bundled capacity is standard capacity product, consisting of the correspondent entry and exit capacity on both sides of every interconnection point. This capacity will be booked by the same shipper on both sides of the IP. Unbundled capacity is a product of capacity available only on one side of the IP, booked only with the TSO who is offering it.

---

[89]Regulation 1775, Art. 18(3).

[90]CAM NC, Art. 3(4).

[91]CAM NC, Art. 6.

[92]The European Federation of Energy Traders (EFET), An EFET Position Paper: Advancing the EU Internal Energy Market: Sector Priorities for the Juncker Commission, 12.11.2014, availabe at: http://www.efet.org/.

6.5.3 The Incremental Capacity Proposal

The 'Incremental Capacity Proposal' aimed to attract efficient and financially-viable investments in new cross-border pipeline infrastructures. The new amendment to the CAM NC offers an Alternative Allocation Mechanisms procedure, akin to open season, as an alternative to the default integrated auction method for allocating incremental capacity described in the CAM NC.

The Alternative Allocation Mechanisms procedure is a standardised procedure for market participants to indicate, in a non-binding way, their demand for incremental capacity. It is a two-step process. First, it enables identification of actual market demand, and second, it allocates capacity on a transparent and non-discriminatory basis. In other words, the decision to invest in incremental capacity is contingent upon commitments from shippers to book enough capacity upfront to fulfil a 'market test'. This is known as market-based investment.

The criteria for eligibility to participate at the open season procedure are made public upfront. In principle, it shall be applied in cases where the pipeline crosses more than two Entry–Exit systems and the demand assessment could not effectively indicate required offer levels. The TSOs and NRAs are to decide on this assessment and approve open season procedures for projects which are unlikely to pass the economic test at the reserve price with a 10-year booking horizon. Open season procedures could be used to allocate all the capacity needed to realise a single project and to avoid inconsistent decisions by different NRAs along the pipeline corridor.

In the course of stakeholder meetings leading to the adoption of CAM NC in 2017, there were other feasible proposals, such as employment of integrated auctions at all entry points along with the provisions of the CAM NC. The CAM NC also provides a market test, as the bids submitted in the auctions help the TSOs determine the need for additional capacity. The bids should provide sufficient information to trigger the investment. However, allocating capacity based on auctions may not be financially feasible for large-scale projects. This might be the case where, for example, two or more entry–exit systems and TSOs are involved or where a 15-years period for auctions, actually 10 years after construction, is not considered to be sufficient to pass the economic test at the reserve price.

Another proposal made during the stakeholder discussions was a coordinated open season approach to resolve the mandatory third-party access problem. It simply means that if you have the capacity and the demand you cannot discriminate between those asking for capacity. This norm is based on the assumption that there is always a capacity deficit that needs to be allocated. The development of the capacity could be arranged in a way that there is no deficit. It is possible to develop a mechanism to create as much capacity as needed by the request for this capacity in the long-term. First, the demand should be articulated in advance and known to the investor; then it is possible to arrange the investment in such a way that there is no deficit. If a deficit still occurs on a seasonal basis, capacity allocation such as via auctions may be organised. It was argued that this coordinated open season approach would significantly diminish deficit.

In its current form, CAM NC provides that incremental capacity is by default allocated by auction process in order to guarantee the highest level of transparency and non-discrimination. However, for large and complex projects, an Alternative Allocation Mechanisms could be allowed on a case-by-case basis decided by the TSOs and approved by the NRAs. This is to provide some flexibility for cross-border investments where there is significant demand. The Alternative Auction Mechanisms should still be aligned among different TSOs and a higher quota of capacity to be set aside for short-term bookings.

6.5.4 Congestion Management

Contractual congestion occurs where capacity at an interconnection point is requested while it is fully booked by existing contracts. In this case, unused capacity should be reallocated to fulfil the pending requests to ensure that capacity is efficiently used. This is the purpose of the CMP. In order to handle contractual congestion, CMP have been amended in the Gas Regulation in 2012[93] providing many measures for CMP, long-term Use-It-Or-Lose-It (UIOLI), Firm Day-Ahead UIOLI, Oversubscription/Buy-Back, Capacity Surrender and finally Secondary Markets for unused capacity,[94] although a formal secondary market for transport capacity does not hitherto exist in all Member States. However, in its 2015 Implementation Monitoring Report on CMP, ACER reported incomplete implementation and limited application of CMP provisions by TSOs across the EU.[95] Things did not improve significantly in 2018, and there was even an increase in contractually congested interconnection points compared to 2017. In response, ACER recommended a revision of the CMP Guidelines to enhance the effectiveness of the measures and improve data reliability.[96]

6.5.5 Tariffs

In an effort to harmonise transmission tariff structures and to allocate costs for incremental capacity within the applicability zone of the Third Energy Package,[97] a Network Code on Tariffs has been developed by the ENTSO for Gas. The network code on rules regarding harmonised transmission tariff structures for gas entered into force on 6 April 2017.[98]

---

[93]Commission Decision of 24 August 2012 on amending Annex I to Regulation (EC) No. 715/2009 of the European Parliament and of the Council on conditions for access to the natural gas transmission networks (2012/490/EU).

[94]Regulation 715, Annex I, guidelines on congestion-management procedures in the event of contractual congestion, para. 2.2.

[95]Implementation Monitoring Report on Congestion Management Procedures in 13.01.2015, ACER, available at: http://www.acer.europa.eu/.

[96]ACER, 2019 Annual Report on Contractual Congestion at Interconnection Points, 27.05.2019, available at: https://www.acer.europa.eu/Official_documents/Acts_of_the_Agency/Publication/Congestion%20Report%206th%20ed_27052019_FINAL.pdf.

[97]The Energy Community is also committed to apply the TEP, but each NC has to be approved separately and to be transposed in national law.

[98]Commission Regulation (EU) 2017/460 of 16 March 2017 establishing a network code on harmonised transmission tariff structures for gas, C/2017/1657, OJ L72.

The NC on Tariffs foresees that the price payable for capacity products is calculated as the sum of its reserve price and, if any, the auction premium.

The main issue with new and incremental capacity is whether shippers should pay for capacity through fixed or floating tariffs. Fixed tariffs may be preferable for the shippers as it would provide greater predictability in long-term commitments. However, the Framework Guidelines for Tariffs, drafted by ACER, suggested that the reserve price is identified as a floating price.[99] Under the floating payable price, where capacity is bought for a gas year beyond the next, the reserve price is not known to network users. Floating tariffs imply that if the revenues of the TSO, at a particular time, are less than expected, and the missing revenues should be compensated for via a tariff increase for the following year. This creates an adverse situation for the shippers, where the actual prices for already-booked capacity could increase, as the final tariff is not set until after the auctions have taken place. However, it would be preferable for the TSO as it allows them to basically review the tariffs every year on the basis of the market environment.

6.5.6 Balancing

The TSO has to undertake an effective balancing mechanism to offset the gap between commercial and physical gas networks. In this regard, a Network Code on Gas Balancing of Transmission Networks was published in 2014 to be applicable as of October 2015.[100] It sets out a vision of market-based balancing at a single virtual trading point inside an entry–exit system. In 2019, the full implementation of this Network Code has not yet been achieved, with countries like Bulgaria, Romania and Greece being significantly behind (due to lack of conditions for

---

[99]Framework Guidelines on rules regarding harmonised transmission tariff structures for gas, 29 November 2013, prepared by Agency for the Cooperation of Energy Regulators (ACER).

[100]Commission Regulation (EU) No. 312/2014 of 26 March 2014 establishing a Network Code on Gas Balancing of Transmission Networks.

market based balancing), and most countries showing some degree of incompliance or inconsistent implementation.[101] Going forward, the conditions in each local market needs to be continuously monitored and evaluated to determine how best to deliver the best value for the end users through improvement of the network codes to enable their full implementation.

6.6   The Role of the Network Codes in Future Infrastructure and Market Design

The Network Codes aim to bring different market zones in the framework of the Third Energy Package under harmonised, transparent, operational and commercial rules. The GTM has been successful to a large extent, however some structural problems and weak competition in the market still persist.

In 2019, ACER identified the main issues as severe in certain regions in the EU where competing sources of supply and new infrastructure are often not heavily utilised as most of the investment focused on security of supply purposes (for instance PCI projects). The reason for insufficient liquidity is also a result of administrative and legal requirements (licensing, security of supply obligations) or exemptions (e.g. from reverse flow requirements).[102]

In order to avoid public funds being utilised for uncommercial projects of a political nature, creating and maintaining an enabling market for private investments in gas infrastructure is necessary. The long-term market uncertainties, regulatory and political, caused by a range of complex interactions in a changing energy market in the EU including

---

[101]ACER Report on the implementation of the Balancing Network Code (3rd edition), 06.08.2018, available at: https://www.acer.europa.eu/Official_documents/Acts_of_the_Agency/Publication/ ACER%20Report%20on%20the%20implementation%20of%20the%20Balancing%20 Network%20Code%20(Third%20edition).pdf.

[102]ACER, Public Consultation on the Bridge Beyond 2025 (Consultation period: 23 July 2019–1 September 2019), available at: https://www.acer.europa.eu/Official_documents/Public_consulta- tions/PC_2019_G_06/The%20Bridge%20beyond%202025%20-%20PC_2019_G_06.pdf.

its transition towards decarbonisation however, carry the risk of making shippers reluctant to commit to long-term capacity bookings. Therefore, the implication of the implementation of the Network Codes on investment decisions is of paramount importance yet remain unexplored.

It can be argued that the GTM offers a solution for the existing third-country long-term supply contracts and has not meaningfully involved the external suppliers in the debate towards its establishment. For instance, even today there is no possibility of non-EU/Energy Community member NRAs taking part in ACER, nor express implication of the regulations on their investments within the EU and this is not likely to change in the near future.

At the time of writing, four of twelve potential Network Codes have been finalised, becoming operational between 2015 and post-2017. Issues are still identified and would be fixed through amendments, with the objective of looking for a balance between the predictability provided by the rules and the flexibility needed to address the process of change.[103] The biggest issue is expected to come from the determination of gas tariffs as more and more long-term contracts expire and are replaced with shorter term horizons, which are often higher than long-term tariffs. This may discourage TSOs from investments in infrastructure and capacity expansion as it poses a barrier to revenue recovery, and as a result could impede trade and further market integration. This is important for future flexibility of the gas system and its decarbonisation, and puts sector coupling between electricity and gas markets to risk as a result of differences between gas and electricity tariff rates. Sector coupling is further discussed below; however, its effective functioning would require competition between power to gas, gas from storage, LNG or electricity storage, which would ideally deliver the benefits to the consumers. Finally, these different tariff levels may potentially provide disincentives for investment decisions in green hydrogen and storage facilities.

---

[103]Gokce Mete, The EU Network Codes and Prospects of Cross-Border Natural Gas Pipeline Projects, in R.J. Heffron and G. Little (eds.), *Delivering Energy Law and Policy in the EU and US: A Reader* (UK: Edinburgh University Press, 2016).

In view of the EU's 2050 strategy and the changing dynamics of the energy sector as a result of the energy transition several regulatory improvements will be necessary, in particular on the governance of the current legal architecture and decision-making framework on infrastructure design. For instance, increased interaction between the electricity and gas sectors may require including more actors on the selection process for investment needs, including industry members from European and third country companies active in the EU, as well as end users so that decisions are not influenced predominantly by political factors but engage technological innovations, local needs and demographic and economic contracts as well as improvements in demand side mechanisms. The 2018 Governance of the Energy Union and Climate Action Regulation[104] obliges EU Member State to produce National Energy and Climate Plans climate by 1 January 2019, covering the 2021-2030 period. These integrated plans are ideally suited to drive the identification of future needs and could play a role in determination of pathways for the facilitation of renewable and low carbon gases.

6.7 Energy Union

Alongside the market architecture endeavoured to be created over two decades under the abovementioned sector packages, the EU is also building an Energy Union. Since its establishment in 2015, the Energy Union, among others, have the objective to promote access to affordable energy, effective regulation and increased competition, building investor confidence, identification of long-term energy needs and exploration of advanced energy technologies. It is a policy tool which also

---

[104]Regulation (EU) 2018/1999 of the European Parliament and of the Council of 11 December 2018 on the Governance of the Energy Union and Climate Action, amending Regulations (EC) No. 663/2009 and (EC) No. 715/2009 of the European Parliament and of the Council, Directives 94/22/EC, 98/70/EC, 2009/31/EC, 2009/73/EC, 2010/31/EU, 2012/27/EU and 2013/30/EU of the European Parliament and of the Council, Council Directives 2009/119/EC and (EU) 2015/652 and repealing Regulation (EU) No. 525/2013 of the European Parliament and of the Council, 21.12.2018 OJ L328.

works towards bringing about a transition to a low-carbon, secure and competitive EU economy. The EU Energy Union furthermore pursues removing market barriers, terminating isolation of small energy islands and the coordination of national policies. The underlying motivation from an energy security perspective is to reduce the vulnerability of Member States and their citizens' to external shocks, interruptions and disruptions.[105] It is notable that it followed the stress test conducted in the summer of 2014 which showed that a severe disruption of gas supplies from the east (i.e. Russia) would still have a major impact throughout the EU. Therefore, the Energy Union recognises the importance of interconnectors, reverse flows and gas storage, the complexity of the energy infrastructure and the security of transit routes, such as the Southern Gas Corridor.[106] The underlying philosophy that is repeatedly stressed is that of working together on security of supply under the principle of solidarity.

The Energy Union is also about ensuring the EU's energy security through long-term planning. Indeed, the State of Energy Union Communication[107] expressed concern that, *"currently only around a third of Member States have comprehensive energy and climate strategies in place beyond 2020"*. The UN Agencies also acknowledge substantial long-term planning and the need for informed choices required in the design of transportation pipelines and transmission grids, particularly for ensuring an infrastructure that is durable and resilient to climate risks.[108]

---

[105]Communication from the Commission to the European Parliament and the Council on the short term resilience of the European gas system Preparedness for a possible disruption of supplies from the East during the fall and winter of 2014/2015 Brussels, 16.10.2014, COM (2014) 654 final.

[106]The Southern Gas Corridor is a term used to describe the European Commission's 2008 initiative to bring gas supplies from the Caspian and Middle East regions to Europe (Second Strategic Energy Review—An EU Energy Security and Solidarity Action Plan" (COM/2008/781)). Today, it denotes a planned infrastructure project, which will make a 3500 km journey from the Caspian Sea to Europe, crossing seven countries, and aimed at improving the security and diversity of the EU's energy supply.

[107]Communication from the Commission to the European Parliament, the Council, the European Economic and Social Committee, the Committee of the Regions and the European Investment Bank, State of the Energy Union 2015, Brussels, 18.11.2015, COM (2015) 572 final.

[108]Sixty-ninth session, item 19 (i) of the provisional agenda, sustainable development: promotion of new and renewable sources of energy. Reliable and stable transit of energy and its role in ensuring sustainable development and international cooperation, report of the

Innovation, technology, competitiveness and research and development are elements of energy security that are considered essential in both the Energy Union package and the UN General Assembly decision (UNGA) 67/263.[109] Creating an economic climate favourable for both enterprises, for the flow of investments and technologies as well as technology transfer, is one of the core principles of the International Energy Charter (IEC) endorsed by the EU along with 90 other states around the globe.[110] The UN Development Programme (UNDP) is confident that universal access to modern energy services is achievable by 2030 and calls on governments to mark this objective as a priority.[111] The Sustainable Energy For All (SE4All) objectives also include, among others, supporting technology and implementing transformational strategies and policies.

These international instruments are mentioned here as, while they are largely aligned with the vision of the EU, the relevant EU foreign policy instruments have a tendency to pursue an energy strategy from an intra-EU perspective. Indeed, there was no established specific external action until 2012 when incompatibility and anti-competitive elements in IGAs between an EU Member State and a non-EU gas supplier became an increasingly worrisome concern for the Commission. Since then the need for transparency in gas supply agreements became one of the core elements of the external aspect of energy policy of the Energy Union. Therefore, the Energy Union has been timely to underpin the external dimension of EU's energy policy which focuses on the position of the EU in global energy markets, and within its immediate

---

Secretary-General, 12.08.14 (A/69/309), available at: http://www.un.org/ga/search/view_doc. asp?symbol=A/69/309&Lang=E.

[109]Resolution adopted by the General Assembly on 17 May 2013, 67/263, Reliable and stable transit of energy and its role in ensuring sustainable development and international cooperation, available at: http://www.un.org/en/ga/search/view_doc.asp?symbol=A/RES/67/263.

[110]The 2015 IEC political declaration is available at: https://energycharter.org/process/international-energy-charter-2015/overview/.

[111]This is also underlined as part of Sustainable Development Goal 7 on ensuring access to affordable, reliable, sustainable and modern energy for all, https://sustainabledevelopment. un.org/sdg7.

neighbourhood. The former relates to climate change and energy security and the latter concerns the EU's policy of expansion of the energy *acquis*.

6.8 Expansion of Rules-Based Approach

6.8.1 Energy Charter Treaty

Energy import dependency in the EU is not new. The energy sector is more efficient today, however in 2014 the EU had to import more than half of the energy it consumed, which is slightly higher than its energy dependency in 1990s (with fossil fuels still being the main source of energy). The EU's dependence on fuels originated from countries outside the EU instigated the need to also establish an international platform and a regulatory space between the EU and potential third-country suppliers, as well as important energy corridors. With this in mind, key European countries initiated signing of the ECT in 1994.[112] The ECT is the first multilateral treaty in the energy sector covering investment protection, transit, trade and environmental aspects of energy. The ECT's constituency originally encompassed, inter alia, most of the energy-rich former Soviet Union and CEE countries.[113] All EU Member States individually and the EU as a whole also signed the ECT. This is done by inclusion of a Regional Economic Integration Organization Clause (REIO) to the Treaty. A REIO clause entails some exceptions to the international treaty in question, to refrain third countries from automatically enjoying the liberalisation advantages of the regional integration formation and to reserve room for legislative manoeuvre for prospective measures to be taken within REIO

---

[112]ECT was opened for signature in Lisbon 17.12.1994 and entered into force on 16.04.1998. It evolved through a non-legally binding political declaration of the European Energy Charter Conference. See Energy Charter's home page available at: http://www.encharter.org/.

[113]Today the ECT has 54 contracting parties. For more information on signatories available at: http://www.encharter.org/index.php?id=6.

that might run contrary to the national treatment principle embodied under the international framework.

The ECT entered into force in 1998. The conclusion phase of the ECT coincided with the development of the first EU energy legislation, the abovementioned 1996 Electricity Directive and the 1998 Gas Directive. Despite its original role as a purely European regional initiative and framework, the ECT gradually expanded into other regions, and became a truly global multilateral instrument. Currently, the Treaty has 54 contracting parties, plus the EU, under the capacity of REIO,[114] and is ratified by 48 countries located in the Central Asian, European and Transcaucasian energy transport routes, including not only producer and consumer countries but also transit States.[115] Notably, States that are not yet WTO members, such as Azerbaijan, Uzbekistan and Kazakhstan[116] are signatories to the ECT.[117]

The ECT provisions focus on four broad areas:

i. Protection of foreign investments based on the extension of national treatment, or most-favoured nation treatment (whichever is more favourable) and protection against key non-commercial risks;

ii. Non-discriminatory conditions for trade in energy materials, products and energy-related equipment based on WTO[118] rules, and provisions to ensure reliable cross-border energy transit flows through pipelines, grids and other means of transportation;

---

[114]According to Art. 1(3) ECT, "[a] *'Regional Economic Integration Organization' means an organisation constituted by states to which they have transferred competence over certain matters a number of which are governed by this Treaty, including the authority to take decisions binding on them in respect of those matters"*.

[115]The Russian Federation signed the ECT and applied it provisionally until the 18 October 2009 inclusive. Detailed information on members and observers of the ECT can be found at: http://www.encharter.org/index.php?id=61.

[116]Azerbaijan, Uzbekistan and Kazakhstan have observer status and are currently negotiating their accession to the WTO.

[117]Notably, Afghanistan became a member of the ECT in 2013.

[118]WTO Agreement: Marrakesh Agreement Establishing the World Trade Organization, 15 April 1994, 1867 U.N.T.S. 154, 33 I.L.M. 1144 (1994). More information on energy services covered by the WTO can be found at: https://www.wto.org/english/tratop_e/serv_e/energy_e/energy_e.htm.

iii. The resolution of disputes between participating States, and, in the case of investments, between investors and host States; and

iv. The promotion of energy efficiency and attempts to minimise the environmental impact of energy production and use.

The ECT, managed by its implementation institute the Energy Charter Secretariat based in Brussels, Belgium, was initially an instrument to expand the market principles of the EU's energy *acquis* to countries outside the EU. This aspect of the Treaty is less relevant now as regulation and policy on energy in the EU moved much faster within the regional block. For almost three decades the Energy Charter has served as a proactive and pragmatic facilitator of cross-border energy cooperation across Europe and the globe. However, the energy system economic and geo-political situation and most importantly the ecology is vastly different today in contrast to the 1990s. The ECT has great potential to further contribute to promoting sustainable energy at the global level and to strengthening global energy security by extending the application of its legal framework to an increasing number of countries. Further, it serves as a very unique investment instrument which sets standards and assures the highest levels of investment protection in the energy sector. It is surprising, then, that the ECT provisions have hardly been revised since its establishment in the 1990s.

There is a growing and common acknowledgement not only by its contracting parties but among the international community that the ECT needs to provide stronger provisions on sustainable development, including on climate change and the clean energy transition, in line with the Paris Agreement. According to an important number of its Contracting Parties, in the area of investment protection, ECT rules do not correspond to modern standards as these standards have been revisited recently. Therefore, in its meeting in Ashgabat in 2017, the Energy Charter Conference opened discussions on potential ECT modernisation. The list of topics for modernisation covers, inter alia, investment protection provisions and related definitions, but also pre-investment commitments, transit, the economic integration agreements clause and some provisions related to dispute resolution. The start of the negotiations to modernise the ECT will ultimately be decided by the Energy Charter Conference. It is expected that this will take place in 2019.

Arguably, the ECT could also play a role in the facilitation of cross-border hydrogen transport and would be an ideal platform for the development of model contracts to standardise the legal issues, similar to its Model Cross-Border Pipeline Agreements, that are also currently being revised to integrate climate change standards post-2015.[119]

6.8.2 International Energy Charter

Since the publication of the Energy Union package in February 2015, the EU has actively promoted and engaged in all foreign energy policy tools. As such, the European Commission signed the IEC in The Hague together with more than 70 other countries including China and Iran in May 2015. This means that today, the ECT framework could be expanded to other regions where new producers and transit States may contribute to the EU's energy security.

While the IEC, a declaration of political intention aiming at strengthening energy cooperation between the signatory states which has now been endorsed by over 90 states and international organisations, is not the obvious instrument for climate change mitigation, it reflects some of the most topical energy challenges of the twenty-first century, including the development and liberalisation of international trade in energy, the development of efficient energy markets, the promotion and protection of energy investments, access to and development of energy sources, nuclear safety, energy efficiency and environmental protection. By including all these relevant issues, the IEC promotes mutually beneficial energy cooperation among nations for the sake of energy security and sustainability, thus fitting nicely into the current global policy agenda.

The IEC broadened the geographic scope of the Energy Charter as more energy producing, consuming and transit countries across

---

[119]Energy Charter Model Agreements for Cross-Border Pipelines could be accessed at: https://energycharter.org/what-we-do/trade-and-transit/model-agreements/.

the world joined as observers. The inclusion of the IEC in the external dimension of the Energy Union presents several advantages. First, it allows the relations between EU energy *acquis* and the ECT to be addressed and clarified. Second, the common foundations and objectives of the EU energy market and the ECT can be streamlined. And third, the experience and lessons learnt in EU regional integration is certainly valuable for processes of regional integration that are taking place across the world and increasing the constituency of the Energy Charter.

6.8.3 Energy Community

The Energy Community, established in 2006 with its Secretariat based in Vienna (Austria) not only aims to extend the rules to non-EU member countries, but also extend the EU internal energy market to South East Europe and beyond on the basis of a legally binding framework. Its foundation, the Treaty Establishing the Energy Community (EnCT)[120] was signed between the EU and nine countries, namely Albania, Bosnia and Herzegovina, FYR of Macedonia, Moldova, Montenegro, Serbia, Kosovo and Ukraine in 2005 and entered into force in 2006. The EnCT, inter alia, aims to create an integrated legal mechanism for energy cooperation among its participants. Both the ECT and the EnCT are concurrently applicable in several Eurasian States.[121] Integration of the markets of the EU's immediate neighbours into the EU single energy market also serves the function of easing access for the EU to their energy resources.

---

[120]Treaty Establishing the Energy Community (2015), available at: https://www.energy-community.org/legal/treaty.html.

[121]Concurrent Members include, GATT & ECT: Turkey, Ukraine, Georgia, Tajikistan, each EU Member State and the EU, EnCT & ECT: Ukraine, Moldova, Bosnia Herzegovina, Albania and the EU, EnCT & WTO: Ukraine, Moldova, Montenegro, Albania and the EU, EU acquis (TEP): EU Member States and potential third countries where the legislation may be applicable.

The EnCT is an international treaty devised to create an integrated regulatory framework for energy cooperation among its participants.[122] Its objective is to provide a secure supply of energy by creating a stable legal environment capable of attracting investment necessary for reliable access to energy sources, for instance, petroleum produced in the Caspian Region, Middle East and North Africa.[123] Some of these States are candidates for EU membership while a few others have no prospect of becoming an EU member, such as Ukraine. Individual EU-member States are not contracting parties per se. However, they may request to be a participant, which would enable them to be represented in EnCT institutions and take part in discussions.[124]

The obligation to follow the EU *acquis* is not the only component of the EnCT. Among other commitments, parties are required to ensure cooperation between them and eliminate discriminatory behaviour within the Energy Community[125] and in case of any supply disruption, to work on a prompt solution.[126]

As mentioned above, the EnCT extends the *acquis communautaire* to the territories of the non-EU Member State contracting parties as adapted to the institutional framework of the Energy Community, which resembles EU institutions, and to the specific circumstances of each contracting party, such as political, legal or economic development levels.[127] The Energy Community *acquis* encompasses the core EU energy legislation in the areas of electricity, gas, environment, competition and renewables. By signing the EnCT, the parties agree to implement the EU *acquis* and to establish a single mechanism for 'Network Energy' defined as including, inter alia, cross-border transmission and/or transportation of gas and electricity as well as transmission of crude oil and petroleum products inside the territory of the Energy

---

[122]Further details regarding the membership structure and the functioning of the Energy Community can be found via its home page at: https://www.energy-community.org/.

[123]EnCT Art. 2(1) and Preamble of the EnCT.

[124]EnCT Art. 95.

[125]Art. 6 and 7 of the EnCT.

[126]Art. 44 EnCT.

[127]EnCT Art. 5.

Community, without internal borders.[128] In line with the EU's new regulatory architecture, this definition represents the merging of cross-border 'energy transit' and domestic flows of energy products within the EnCT's constituency.

At the time when EnCT was first established, the second Gas Directive and Regulation 1775 were still in force. The implementation commitments of the then non-EU contracting parties were based on the relevant legislation. However, the EnCT paves the way for amendments to the treaty in line with the evolution of EU law.[129] Hence, the Energy Community adopted a decision in 2011 to implement the Third Energy Package legislation, including provisions on OU, access regimes and implementing rules such as the Network Codes, by 1 January 2015.[130] In 2019, the implementation of the legislation continues with an average implementation score of 43% (the score varies among different countries). Implementation of the gas sector regulations moves at a slower pace than for electricity The focus is currently on continuance of unbundling and the certification of TSOs. The Energy Community Secretariat believes that in its constituency, the energy transition has two steps. First, liberalisation and opening up markets (transitioning away from the post-socialist energy sectors) and second, decarbonisation; the former is an enabler of the latter.[131]

This is important, as a day after the adoption of the 2050 Long-Term Climate Strategy by the EU, the Energy Community contracting parties agreed to establish three distinct 2030 energy and climate targets for GHG emissions reduction, energy efficiency improvements and increasing the share of renewable energy. This political agreement, underpinned by the General Policy Guidelines, foresees that these targets will be in line with the EU's 2030 targets taking into account relevant

---

[128]Art. 3(a) and (b) EnCT.

[129]Art. 25 EnCT.

[130]Decision on the implementation of Directive 2009/72/EC, Directive 2009/73/EC, Regulation (EC) No. 714/2009 and Regulation (EC) No. 715/2009 and amending Articles 11 and 59 of the Energy Community Treaty.

[131]Energy Community, State of Implementation Report 2018, available at: https://www.energy-community.org/implementation/IR2018.html.

socio-economic differences, technological developments and the Paris Agreement on Climate Change. This being said, due to the heavy subsidisation of coal in the Western Balkan countries in particular, the speed of decarbonisation of these countries has been slow. In 2019, the Energy Community reported coal still represents 97% of electricity generation in Kosovo, 70% in Serbia and Bosnia and Herzegovina, more than half in North Macedonia, around half in Montenegro and 28% in Ukraine. The commissioning of the TANAP and TAP pipelines and further interconnectors may open options for a coal-to-gas switch in the region, with emphasis on hard to decarbonise sectors of the respective economies, with significant gains in GHG emissions in the short-term alongside ramping up of renewable energy investments including biomass from agricultural waste (for heating). In the long-term, hydrogen-based heating and power generation solutions may offer promising applications, which can be traded within the Energy Community single market. However, this would only come after the necessary market conditions and legal framework for these technologies are well established in the EU, perhaps under the 2020 Gas Package.

6.9   International Agreements

During the 1990s, individual Member States also signed Bilateral Investment Treaties (BITs) with third-countries (alongside the ECT framework) including Eastern European countries before their accession to the EU (which later became intra-EU BITs). These BITs provide an investor from the third country with protection against discriminatory acts of host governments who may adversely affect the investment, for example, through expropriation. More often than not, BITs also contain investor-State investment arbitration clauses which give foreign investors the right to initiate a dispute settlement with host States.[132] Moreover, some Member States signed IGAs with producing

---

[132]Press release, The European Commission, EU agrees rules to manage investor-state disputes, 28.08.14, available at: http://europa.eu/rapid/press-release_IP-14-951_en.htm.

third-countries such as Russia for cross-border energy supplies and the construction of energy infrastructure.

Since 2003, a number of new pipeline agreements in the form of IGAs and MoUs were signed across Eurasia including towards implementation of the Southern Corridor initiative. Cross-border pipelines operate between different legal and regulatory regimes, and agreed stable, legal and fiscal regimes can be created through HGAs, IGAs, long-term transportation agreements and a suitable tariff structure for transit countries. The project owner (investor) is not usually a party to an IGA and does not, therefore, have standing to enforce that agreement against the various governments if any one or more of them fail to comply with their obligations. Thus, the regulatory framework of the project is ensured and government warranties to project investors are guaranteed through these agreements.

These BITs and IGAs have for some time been agreed by EU members between members and among members and third countries in parallel with the creation of an internal energy market. Such IGAs in various parts of the EU either establish common rules and procedures for specific energy projects or regulate interactions in the energy sector between the States involved. However, since the 2012, the information exchange mechanism was adopted[133] in the interests of coordinated action and cooperation in relations with external energy producers as well as ensuring compliance with EU law in relation to energy deals with non-EU suppliers.

Member States were to notify the Commission of IGAs related to the purchase, trade, sale or supply of energy and construction of energy infrastructure signed between them and third countries which has an impact on the internal energy market or security of energy supply in the EU. The Commission then would evaluate those agreements for compatibility with EU law. Four years following implementation of

---

[133]Decision No. 994/2012/EU of the European Parliament and of the Council of 25 October 2012 establishing an information exchange mechanism with regard to intergovernmental agreements between Member States and third countries in the field of energy.

this decision, a new legislative proposal, the so-called 'winter package' as part of the action plans of the Energy Union was published in February 2016.[134]

The winter package included, inter alia, a proposal for the revision of the abovementioned 2012 decision on information exchange for energy related IGAs with third countries and revision of the 2010 Gas Security of Supply Regulation. According to the recommended rules, the Commission would be able to see the IGAs before they are signed, as it is difficult to re-negotiate them afterwards. The Member States response to the new proposals were somewhat divergent however, as there was a tendency to refrain from conceding more powers to the Commission on security of supply. The position of the Council and the Parliament therefore differed regarding some core issues of the proposal. Nevertheless, an agreement was reached on 7 December 2016. On 16 December 2016, the Council's Permanent Representatives Committee endorsed the text of the final compromise. The decision was adopted on 5 April 2017 and became the first legislative dossier adopted under the Energy Union strategy. The revised IGAs decision introduces a compulsory *ex ante* compliance check with the EU legislation of IGAs on gas and oil, conducted by the Commission.[135] The compromise is that IGAs related to electricity will only be subject to *ex-post* compliance checks for now but a review clause leaves open the possibility to introduce a compulsory *ex-ante* check at a later stage.

To date, there has been no re-negotiation of any IGA. The only example of the Commission being granted permission to sit in the negotiation of a project agreement from the outset is the mandate adopted in 2011

---

[134]Proposal for a Regulation of the European Parliament and of the Council Concerning measures to safeguard the security of gas supply and repealing Regulation (EU) No. 994/2010 Brussels, 16.2.2016; Proposal for a Decision of the European Parliament and of the Council on establishing an information exchange mechanism with regard to intergovernmental agreements and non-binding instruments between Member States and third countries in the field of energy and repealing Decision No. 994/2012/EU, Brussels, 16.2.2016.

[135]Decision (EU) 2017/684 of the European Parliament and of the Council of 5 April 2017 on establishing an information exchange mechanism with regard to intergovernmental agreements and non-binding instruments between Member States and third countries in the field of energy, and repealing Decision No. 994/2012/EU, OJ L99.

to negotiate a legally binding treaty between the EU (on behalf of all 28 Members of the Union), Azerbaijan and Turkmenistan to build a Trans-Caspian pipeline system.[136] If an IGA would be signed in the future, it would be the first time that the EU be party to an IGA for an energy infrastructure project.

A recent option proposed to bring the North Stream 2 project in line with EU law was to launch an IGA between the EU, Germany, Russia and other countries that the pipeline will cross: Denmark, Finland and Sweden. As stated above, IGAs provide a hybrid regime, balancing non-corresponding aspects of national laws. However, the Legal Service of the EU Council noted that a possible special regime created for North Stream 2 could only cover the operational aspects of the project and not its construction or commissioning. The Commission does not have a legal right to block the construction of such a pipeline.[137] The Legal Service also underlined that this special regime would stem from a political compromise and not a legal necessity.

In principle, the Commission's participation in IGAs with EU-wide energy security implications has a positive aspect. However, this can only be performed when all the states involved consent, including third (producer and transit) countries, in addition to the EU Member States. It is also important that such negotiations are de-politicised and focus on the project's cost-benefit analysis and commercial realities. This is an unlikely scenario for the negotiation of an IGA for the North Stream 2 case (Russia and Germany are unwilling to engage in one—for Germany this is purely a commercial project, signed and negotiated essentially between commercial entities and not governments). Politicisation of the project could set a negative example for future projects involving the Commission at the negotiation table and could even jeopardise the Trans-Caspian negotiations.

---

[136]European Union Press release, EU Starts Negotiations on Caspian Pipeline to Bring Gas to Europe, 12.09.2011, available at: https://europa.eu/rapid/press-release_IP-11-1023_en.htm.

[137]Leaked opinion of the legal service dated 27 September 2018 is available at: http://www.politico.eu/wp-content/uploads/2017/09/SPOLITICO-17092812480.pdf.

The 2016 Revision of the Regulation on Security of Gas Supply[138] also foresees standardised and procedural streamlining and a trend from national to regional plans. The revised regulation was adopted in October 2017, and among others introduces a solidarity principle as a binding legal mechanism to be used as a last resort with fair compensation.[139]

Further, in September 2017 the Commission carried out an Impact Assessment to replace the traditional investor-state-dispute-settlement mechanism, which is the existing method of dispute resolution for over 1400 BITs and the ECT that is in force between EU Member States and third counties.[140] Similar to the ECT, but on a bilateral basis, BITs provide an investor from a third country with protection against discriminatory acts from host governments, which may adversely affect the investment, for example, through expropriation. Often, the BITs also contain an investor-host State arbitration clause which gives foreign investors the right to initiate a dispute settlement with the host State.[141] In its Impact Assessment, the Commission was seeking a mandate to negotiate, with its trading and investment partners, a multilateral court for the settlement of investment disputes on behalf of the Member States.

The independence and effectiveness of the proposed Multilateral Investment Court, provided that such a mechanisms is established, is

---

[138]Proposal for a Regulation of the European Parliament and of the Council concerning measures to safeguard the security of gas supply and repealing Regulation (EU) No. 994/2010, COM/2016/052 final—2016/030 (COD).

[139]Regulation (EU) 2017/1938 of the European Parliament and of the Council of 25 October 2017 concerning measures to safeguard the security of gas supply and repealing Regulation (EU) No. 994/2010, OJ L280.

[140]European Union, commission staff working document impact assessment multilateral reform of investment dispute resolution Accompanying the document Recommendation for a Council Decision authorising the opening of negotiations for a Convention establishing a multilateral court for the settlement of investment dispute, SWD(2017) 302 final, Brussels, 13.9.2017, available at: https://ec.europa.eu/transparency/regdoc/rep/10102/2017/EN/SWD-2017-302-F1-EN-MAIN-PART-1.PDF; European Parliament, Multilateral court for the settlement of investment disputes, Ulla-Mari Tuomine, 24.11.2017, available at: http://www.europarl.europa.eu/think-tank/en/document.html?reference=EPRS_BRI(2017)611016.

[141]European Commission Press release, The European Commission, EU agrees rules to manage investor-state disputes, IP/14/951, 28.8.2014.

yet to be tested. There is a good opportunity for the Commission to find a common denominator with third country suppliers via the negotiation of IGAs and creating a hybrid legal system mutually agreed with all countries concerned. Yet, it should be further elaborated with principles that guarantee some level of non-discrimination *vis-a-vis* investors of different nationalities and tools to depoliticise negotiations. However, the Member States' rights under Article 194(2) TFEU to determine its choice between different energy sources, and the general structure of its energy supply should be respected. Member States should not be deprived of their right to negotiate bilateral IGAs with exporting third countries on their own.[142]

The developments introduced above presents the direction of the Commissions' increased powers in the decision making process regarding energy sector developments of individual Member States (since as early as 1990s) (to a degree that it can sit in on negotiations alongside the Member States and third countries concerned)[143] despite the fact that Article 194 TFEU confers on the Commission the power to take legislative action in the energy field, its second paragraph allows 'energy rights'/opt-outs to persist. This means that Member States can continue to decide on their energy mix.[144] Things went one-step forward with the 2019 amendment of the Gas Directive. This will be discussed next as it granted more powers to political bodies in commercial decision making and added further uncertainties with possible adverse impacts on the markets and current and future projects with gas producing and transit countries.

---

[142]Eurogas, Eurogas views on the modification of the Gas Directive, January 2018, available at: http://www.eurogas.org/uploads/media/18PP002_-_Eurogas_views_on_the_modification_of_the_Gas_Directive.pdf.

[143]More details on the information exchange mechanism for agreements between governments is available at: https://ec.europa.eu/energy/en/topics/international-cooperation/intergovernmental-agreements.

[144]Ernesto Bonafé and Gökçe Mete, Escalated Interactions Between EU Energy Law and the Energy Charter Treaty, *The Journal of World Energy Law & Business*, Volume 9, Issue 3, 1 June 2016.

6.10 Decision Making Framework Under the EU Natural Gas Market Rules: An Altering Journey to Destination Unknown?

6.10.1 2019 Gas Directive Amendment

The 2019 Gas Directive Amendment presents an important change in the decision making framework for pipeline connections with non-EU countries. These amendments followed a 'Quo Vadis EU Gas Regulatory Framework' study initiated by the Commission, which carried out a welfare check and concluded that functioning of the European gas markets and overall EU welfare can be improved through amendment of the current regulatory framework.[145]

The legislative process leading to its adoption was not free from some really heated debates and it raised major legal questions. These changes also resulted in an arbitration proceeding brought against the EU, but this will be discussed later. First, it is important to introduce the events that took place prior to the finalisation of its precise terms. In 2017, the Commission proposed an amendment to the 2009 Gas Directive to extend the provisions of the Third Energy Package to existing off-shore pipelines and new and incremental capacities with third countries which are currently explicitly excluded from the application of this legal framework, as they are covered by the UN Convention for the Law of the Sea (UNCLOS). The Commission proposed identifying these pipelines as 'interconnectors', to bring them under the scope of the Gas Directive and Regulation.[146] This move was reflected in a Letter of

---

[145]European Commission, Quo Vadis EU Gas Market Regulatory Framework—Study on a Gas Market Design for Europe, February 2018, available at: https://ec.europa.eu/energy/en/studies/study-quo-vadis-gas-market-regulatory-framework.

[146]Proposal for a Directive of the European Parliament and of the Council amending Directive 2009/73/EC concerning common rules for the internal market in natural gas, COM (2017) 660 final, 2017/0294 (COD) Brussels, 08.11.2017, available at: https://ec.europa.eu/energy/sites/ener/files/documents/act_gas_dir_adopted.pdf.

Intent drafted by EU President Junker, which proposed common rules for gas pipelines entering the European internal gas market as an initiative to be completed by end-2018.[147]

Within the European Parliament, the proposal was sent to the Industry, Research and Energy Committee (ITRE).[148] The ITRE committee discussed the draft report on the 11 of January 2018, and a number of amendments were proposed.[149] Some Committee members were of the opinion that export pipelines, whether upstream or not, are the means for Member States to exploit their natural resources. They noted that according to Article 194, TFEU prohibits measures that affect the Member States' right to determine the conditions for exploiting their energy resources. Therefore, it is not acceptable and legally justifiable to include them within the scope of the Gas Directive.

The classification of the pipelines in the 2009 Gas Directive and the amendments are problematic. While the Gas Directive provides a definition of direct lines, transmission, distribution, upstream and interconnector pipelines, there is no definition of pipelines that connect the transmission systems of a third country with a Member States transmission system, the so-called export pipelines. The Commission's proposal referred to these lines as interconnectors. However, some members of the Committee contested this, noting that EU legislation could only apply within the territorial jurisdiction of the EU. As for submarine

---

[147]Commission Staff Working Document Assessing the amendments to Directive 2009/73/EC setting out rules for gas pipelines connecting the European Union with third countries, Accompanying the document Proposal for a Directive of the European Parliament and of the Council amending Directive 2009/73/EC concerning common rules for the internal market in natural gas, available at: https://ec.europa.eu/energy/sites/ener/files/documents/swd_-_gas_dir_amendment_-_final.pdf.

[148]Briefing EU Legislation in Progress: Common rules for gas pipelines entering the EU internal market, 23.01.2018, available at: http://www.europarl.europa.eu/RegData/etudes/BRIE/2018/614673/EPRS_BRI(2018)614673_EN.pdf.

[149]Amendments to the draft report by Jerzy Buzek on the Proposal for a Directive of the European Parliament and of the Council amending Directive 2009/73/EC concerning common rules for the internal market in natural gas Proposal for a directive (COM (2017)0660—C8-0394/2017—2017/0294(COD)), Committee on Industry, Research and Energy, 26.1.2018, available at: http://www.europarl.europa.eu/sides/getDoc.do?pubRef=-//EP//NONSGML+COMPARL+PE-616.573+02+DOC+PDF+V0//EN&language=EN.

pipelines, members noted that the Gas Directive shall not apply in the EEZ of Member States as they would be regulated by the relevant provisions of UNCLOS.

On the other end of the spectrum, some Committee members suggested strengthening the proposal by specifying the territorial applicability of the Directive in the main text (and not only in the Preamble). This, in their view, would allow NRAs to fix or approve tariffs or methodologies for pipelines with third countries, thus enabling the Commission to closely monitor the derogation decision given by Member States. Another suggestion was to specify the time limit for derogations, with a maximum of 10 years after the revised Directive comes into force. Some experts outside the EU institutions argued that the Gas Directive amendments are not necessary, as the same results could be achieved by broader interpretation of the existing rules,[150] which is an even more damaging idea. This could let the EU authorities make decisions that could impact third county investors and effectiveness for the domestic market which are not based on law, but rather based on interpretations made by politicians. This would hardly be accepted by stakeholders. Others in the Committee were entirely opposed to these proposals. In their view, the Parliament should reject it completely since the Commission made the proposals without stakeholder consultation or an Impact Assessment on whether the Gas Directive requires a revision.

Indeed, ITRE Committee members noted that the legislative proposal was published without a stakeholder consultation, an impact assessment, or a regulatory fitness check,[151] despite the fact

---

[150]Blog Post by SZYMON ZAREBA, The Gas Directive and Its Application to EU-Third Country Pipelines, Energy Post, 16.10.2017, available at: http://energypost.eu/making-the-gas-directive-apply-to-import-gas-pipelines/?utm_campaign=shareaholic&utm_medium=linkedin&utm_source=socialnetwork&lipi=urn%3Ali%3Apage%3Ad_flagship3_feed%3Bqu0C1jd-VR7SXgfTexZfA%2Bw%3D%3D.

[151]Amendments to the draft report by Jerzy Buzek on the Proposal for a Directive of the European Parliament and of the Council amending Directive 2009/73/EC concerning common rules for the internal market in natural gas proposal for a directive, available at: http://www.europarl.europa.eu/sides/getDoc.do?pubRef=-//EP//NONSGML+COMPARL+PE-616.573+02+DOC+PDF+V0//EN&language=EN.

that these are required under the Better Regulation Guidelines.[152] Furthermore, it was suggested that the proposed text be amended in such a way as to make it a prerequisite for the Member State, at the first point of interconnection, to agree with the third country where the pipeline originates or transits. This would make sense; without the approval of the third country, imposing rules on countries that have not been involved in developing the infrastructure defies business logic, even if the costs of the infrastructure are recovered. There were also reasonable proposals made for the decision making Member State to consult with all the NRAs of all the Member States to which the infrastructure in question is connected.[153]

On the 21 March 2018, the final report which included the integrated comments of the ITRE Committee members, went to a vote. The members voted 41 in favour, 13 against, with 9 abstentions.[154] The amendments were adopted in April 2019 by the Council with a compromise. The entire negotiation exercise, which took over a year, presented the divided opinions of the Member States, with countries in the Baltic and Eastern Europe supporting the changes, and others including Germany (the end destination of North Stream 2) opposing it. The compromise is that in contrast to the original proposal, the final text states that the amended Gas Market Directive would apply only in the *"territory and territorial sea of the Member State where the first interconnection point is located"*. This territorial restriction avoids any conflict with UNCLOS over the operation of a pipeline in the EEZ. The entry point of the interconnection in these cases will have to be a virtual point where the border of the territorial water is located.

---

[152]Better Regulation: Guidelines and Toolbox is available via: https://ec.europa.eu/info/better-regulation-guidelines-and-toolbox_en.

[153]Amendments to the draft report by Jerzy Buzek on the Proposal for a Directive of the European Parliament and of the Council amending Directive 2009/73/EC concerning common rules for the internal market in natural gas Proposal for a Directive, available at: http://www.europarl.europa.eu/sides/getDoc.do?pubRef=-//EP//NONSGML+COMPARL+PE-616.573+02+DOC+PDF+V0//EN&language=EN.

[154]Common rules for the internal market in natural gas: extracts from the vote and statement by Jerzy BUZEK (EPP, PL), Rapporteur and illustration shots, and 21.03.2018, available at: https://multimedia.europarl.europa.eu/en/common-rules-for-internal-market-in-natural-gas_I152643-A_a.

The amendments target not only future but also existing pipelines, such as those connecting the EU with Algeria, Libya, Tunisia, Morocco, Turkey, Russia and post-Brexit, the UK.[155] However, Article 49a of the amendments notes that existing pipelines can benefit from a derogation from, among others, provisions of unbundling, third-party access and tariff regulation. This shall be decided on a case by case basis. The exemption covers only these interconnections of the gas transmission line located in its territory and territorial sea and could only be granted on objective reasons which are exemplified as enabling the recovery of the investment made or due to reasons of security of supply. Neither the exemption decision, nor derogation should have a detrimental impact on competition, effective market functioning or security of supply in the Union. This approach is different from the exemption procedure stipulated in Article 36 of the Gas Directive; in the latter case, derogation is granted in the view of complex legal structures that are already in place and that without the exemption, the project would not be built. The derogation shall be limited in time up to 20 years based on objective justification, renewable if justified and may be subject to conditions which contribute to the achievement of the above conditions. The derogation decision is to be taken by the Member State of first connection. However, without a doubt those decisions will be evaluated by the Commission. In the case that the pipeline is traversing through multiple Member States, the Member State of first connection will discuss with all the Member States concerned when deciding on the derogation (while it is not clear why other Member States should be concerned). The Commission may bring infringement proceedings against the Member State granting the derogation if it fails to comply with the very restrictive conditions of the abovementioned derogation provision. The amendments de facto increases the powers of the Commission in deciding on the fate of cross-border pipelines and potentially limits the powers of Member States under Article 194 TFEU on deciding their own

---

[155]News Article, ICIS, EC Proposes Applying Internal Market Rules to Import Pipes, 15.10.2017, available at: https://www.icis.com/resources/news/2017/11/15/10163918/ec-proposes-applying-internal-market-rules-to-import-pipes/.

energy mix.[156] The differentiated rules and conditions of the derogation (under Article 36 and 49a) creates the risk that different rules will apply to different pipelines within the EU, which would act as a barrier for access to European gas markets.

The Gas Directive Amendment seeks to obtain the same benefits that have been identified in the Security of Supply Regulation, Network Codes and the Energy Union. The objective is to ensure that competition is not distorted and that gas can flow freely and efficiently to wherever it is needed within the EU. It also aims to improve supply security and ensure that competition increases among suppliers importing gas to the internal market. However, it is not clear from the amendment precisely how extending the application of the Third Energy Package to offshore pipelines will contribute to achieving these objectives. Third Party Access could enhance competition among suppliers, but, as the OPAL exemption case showed, if there are no alternative shippers, the capacity can run empty despite the existence of demand which is more detrimental for security of supply.

The justification for the amendment also mentions bringing transparency to upstream segments of the third country pipelines to best serve the interests of the EU customers. Transparency in the entry/exit points at the EU border is already achieved under the Balancing Network Code, and ENTOG's transmission capacity map provides an overview of Europe's main high-pressure transmission lines and provides information on technical capacity at cross-border points.[157] The extension of the Third Energy Package to offshore pipelines means that these entities would have to unbundle. However, in most cases submarine pipelines are constructed under a joint venture model, i.e. a mixture of several companies, such as the North Stream pipeline and the TAP pipeline (although TAP is constructed as merchant exempt—hence it is not subject to the Third Energy Package for 25 years).

---

[156]Kim Talus, EU Gas Market Amendment—Despite of Compromise, Problems Remain, February 2019, OGEL (advance publication).

[157]ENTOG's cross-border transmission capacity map 2017 is available at: https://www.entsog.eu/maps/transmission-capacity-map.

The amendments, adopted rather quickly without proper cost-benefit analysis, carry the risk to further disincentive private investments. There was also a lack of reasonable consultation with the industry stakeholders. The associations were given a very short period of time for public consultation, which also coincided with the Christmas holiday period, making it very difficult to respond in time. The renegotiation of existing IGAs will not only be costly and time consuming but may also result in bottlenecks. The terms and conditions for granting derogations are not clear, such as what constitutes the principles of non-discrimination and competitiveness when applied to the selection of projects. Different arrangements for pipelines across the EU may not add too much to transparency and competition among suppliers and could instead lead to distortions as projects destined for the EU may go to other available markets. All potential gas suppliers of the EU have alternative markets, including Russia. A further issue is that the externalisation of EU energy *acquis* covers only the underwater sections of pipelines, and not LNG, which is not in line with the free and open market principles of the EU. This may result in potentially more expensive LNG having a competitive advantage over cheaper pipeline gas which may be detrimental for end users.

In response to the Gas Directive amendment, North Stream 2 expressed its intention to initiate an ECT arbitration proceeding against the EU in 2019.[158] While Russia has now withdrawn from the Treaty, the company is registered under Swiss laws and Switzerland is a contracting party to the ECT. North Stream 2 claimed that the Amended Directive targeted the pipeline in a discriminatory fashion, and that it was the only pipeline subject to the new law where a final investment decision had already been made. This therefore breached its legitimate expectations protected under the investment regime of the ECT. The letter was intended to serve as a 'notice of dispute' to the EU under ECT rules and asked the EC to respond by 13 May 2019. It has been reported that efforts to reach an amicable settlement has so far failed.

---

[158]Letter from North Stream 2 to the Commission, 12.04.2019, available at: https://trade.ec.europa.eu/doclib/docs/2019/july/tradoc_158069.pd_Redacted.pdf.

The ECT arbitration proceedings can be very costly (as the Yukos case showed)[159] and could cost a considerable amount of tax payer money. Whether or not it represented bad law making, or would derive any significant benefits to the functioning of the EU gas markets is unknown. To explore the subject further, the next section will look at some of the discussions raised in the Quo Vadis report which initially recommended amendments to the Gas Directive.

The Quo Vadis document 'A Study of the Gas Market Regulatory Framework', as an 'intellectual exercise', investigated whether there is a need to amend the current regulatory framework for the gas sector and, if so, how to go about it. The study identified key deficiencies in the internal market and suggested radical changes to the existing regulatory system.[160]

Among other things, the Quo Vadis study rightfully addresses the problem of the imbalanced structure of gas transportation tariffs, the continued physical congestion in the supply chains, contractual mismatches, and inadequacy of reverse flows (bi-directional pipelines). The report also addresses the risk of stranded assets due to the low utilisation level of the EU gas transportation system, strong NRA control and the lack of transparency in the functioning of the markets, as well as technical and network operation ineffectiveness due to cost-inept infrastructure investment decisions. These are all issues that stem from market participants' failures to fully comply with the Third Energy Package, the Network Codes and the Energy Union provisions (and all associated decisions such as the TEN-E and the Security of Supply Regulation). The study, on the other hand, suggests that if the Third Energy Package continues to apply as it is, the shift from long-term to spot contracts will lead to increased pricing among EU Member States.

The Quo Vadis study in the EU also endeavoured to address the issue of return of energy infrastructure investment. The European

---

[159]Yukos Universal Limited (Isle of Man) v. The Russian Federation, UNCITRAL, PCA Case No. AA 227.

[160]European Commission, Quo Vadis EU Gas Market Regulatory Framework—Study on a Gas Market Design for Europe, February 2018, available at: https://ec.europa.eu/energy/en/studies/study-quo-vadis-gas-market-regulatory-framework.

Federation of Energy Traders (EFET) was concerned that there is a risk of an increase in the number of stranded assets in the EU if projects that are non-commercial but constructed to ensure security of supply are not properly accompanied with a cost-benefit analysis.[161] This inevitably puts commercially-viable new projects at risk, as most resources are spent on strategic projects. Stranded assets also increase risks as often the highest demand centres and easiest/shortest routes are selected. Currently, the largest investment gap in the EU in terms of gas infrastructure is, as mentioned several times in the book, in South Eastern Europe, which is facing actual security of supply risk. If, for any reason, all Russian gas transit through Ukraine is interrupted, the region will have to be prepared for re-routing the flows or attracting new sources of gas. The TAP project has therefore been implemented as an exempt merchant pipeline. Looking forward, however, the volumes in TAP are unlikely to meet the demand from this region and other alternatives such as the Turk Stream extension, the Trans Caspian, and East-Med do not look very promising. These are the regions where gas sector decarbonisation could have even more value as it would increase the variety of energy sources available and provide local solutions, in particular with renewable gases.

Further, radical changes suggested in the report include the enlarging or merging of market zones within the internal market to redistribute revenues among TSOs. There is also a recommendation for removal of entry–exit tariffs and instead introducing intra-zone tariff losses to entry tariffs of the merged/enlarged zone or splitting them between wholesale market and retail zones. This means that 50% of TSO revenue losses are reflected to the exporters to the EU, and 50% to the end users.[162] There is also a proposal to establish a TSO Compensation Fund to increase the tariffs so that the TSOs could better finance the maintenance and development of the transmission system.

---

[161]EFET, Greece–Italy Incremental Capacity Proposal, 18.12.2017, available at: http://www.efet.org/Files/Documents/Downloads/EFET%20comments%20incremental%20capacity%20Greece%20-%20Italy.pdf.

[162]European Commission, Quo Vadis EU Gas Market Regulatory Framework Study on a Gas Market Design for Europe, February 2018, available at: https://ec.europa.eu/energy/sites/ener/files/documents/quo_vadis_report_16feb18.pdf, pp. 124–131. Ibid.

Another far-reaching suggested change is moving gas delivery points within the EU to the external border of the energy regulatory border of the EU, which not only includes Member States but also the Energy Community area, which extends to Ukraine, at the border of Russia (the largest single supplier of gas used in the EU). The Quo Vadis study aimed to provide a quantitative modelling exercise without considering the enforceability of the scenarios.[163] Indeed, the reaction from industry stakeholders reflected that there are serious concerns about the practicability of these models. For instance, EFET noted that "*Quo Vadis should start with an assessment of the existing shortcomings identifying those which are likely to be addressed by proper implementation of the existing regulatory framework*".[164] In relation to the proposed tariff reform and the market mergers scenarios, EFET criticised the study for not considering different pricing options and not considering that the regulation needed to "*guarantee a level playing-field for different supply routes to Europe, bearing in mind that the EU has a domestic production of gas insufficient to cover its needs*".[165] Most importantly, EFET highlighted that "*tariffs should be able to attract import flows, particularly from those third countries that increase the diversification of sources*".[166]

Gas Infrastructure Europe (GIE) condemned the exercise as they were given very little time to reflect on the report before it was finalised. Addressing the Commission, GIE suggested that "*the Commission more clearly expresses which kind of problems a revised market design should solve*".[167] GIE recommended the Commission to first focus on achieving "*a well interconnected, integrated and flexible gas infrastructure*

---

[163]News Article, Natural Gas World, EU Quo Vadis: A Theoretical Exercise with an Anti-Russian Flavour, 19.10.2017, available at: https://www.naturalgasworld.com/gpp-eu-quo-vadis-a-theoretical-exercise-with-an-anti-russian-flavour-56079.

[164]EFET, Comments to the Quo Vadis Modelling Work, 18.08.2017, available at: http://www.efet.org/Files/EFET%20comments%20to%20the%20QuoVadis%20modelling%20work.pdf.

[165]EFET, Comments to the Quo Vadis Modelling Work, 18.08.2017, available at: http://www.efet.org/Files/EFET%20comments%20to%20the%20QuoVadis%20modelling%20work.pdf.

[166]EFET, Comments to the Quo Vadis Modelling Work, 18.08.2017, available at: http://www.efet.org/Files/EFET%20comments%20to%20the%20QuoVadis%20modelling%20work.pdf.

[167]GIE response to the invitation for written comments on the "Quo Vadis EU Gas Market Regulatory Framework—Study on a Gas Market Design for Europe" preliminary report presented on 26.06.2017, available at: https://www.gie.eu/index.php/publications/cat_view/2-gie-publications.

network in Europe as a way to enhance supply security, integrate European wholesale markets and ensure the free flow of energy across the borders" and facilitating gas to remain competitive in an electricity balancing market.[168] The GIE criticised the study for not taking into account some innovative uses of gas, such as renewable gases and hydrogen, and failing to ensure that there is no discrimination between flexible providers among all gas infrastructure assets. According to GIE, there are more pressing issues to be fixed within the internal market, such as the amount of gas storage facilities, for which the Commission is tasked to ensure high withdrawal rates, that are available for flexibility and security of supply purposes.

According to numerous stakeholders, then, regulatory intervention should be avoided.[169] As noted by EFET, "*modelling should not completely replace qualitative analysis based on 'real world' situation (observed pricing strategies and market inefficiencies; unequal implementation of EU gas market legislation, etc.)*".[170] The Third Energy Package is perceived to have already contributed to harmonising market arrangements and incentivising gas portfolios.[171] What remains to be fixed, and what the Commission, in an ideal world, is trying to achieve, is a universal application of the Network Codes and the energy *acquis* as a whole. Some of the Network Codes have just been entered into and there will be some challenges to their implementation as we move forward, which

---

[168]GIE response to the invitation for written comments on the "Quo Vadis EU Gas Market Regulatory Framework—Study on a Gas Market Design for Europe" preliminary report presented on 26.06.2017, available at: https://www.gie.eu/index.php/publications/cat_view/2-gie-publications.

[169]Europex feedback to Part 1 of the "Quo Vadis EU Gas Market" Study, 10.07.2017, available at: https://www.europex.org/position-papers/europex-feedback-to-part-1-of-the-quo-vadis-eu-gas-market-study/.

[170]EFET, Comments to the Quo Vadis Modelling Work, 18.08.2017, available at: http://www.efet.org/Files/EFET%20comments%20to%20the%20QuoVadis%20modelling%20work.pdf.

[171]EFET response to the decision to Commission the "Quo Vadis" Consultants' Study on Gas Market Design, 16.01.2017, available at: http://www.efet.org/Files/Documents/Gas%20Market/Tariffs/EFET%20Quo%20Vadis%20Response.pdf.

the Commission may consider as a priority area. EUROGAS, representing the European gas wholesale, retail and distribution sectors, also noted that implementation of the Third Energy Package should remain a priority.[172]

The Energy Union initiative provided an opportunity to correct inadequacies without significant regulatory intervention, which is, as a rule, detrimental for legal predictability and is at odds with the open market principle. However, the Commission started translating the Energy Union concept into legislation with the revised Security of Supply (2017) and Information Exchange legislations (2017), and with its proposal to amend the Gas Directive (2018). The 2020 Gas Package will follow this exercise. In principle, the Energy Union is about pooling resources, connecting networks, diversifying energy sources, supply routes and suppliers, and at the same time making EU energy more efficient, less dependent on exports and also leading on renewable energy technologies. The Energy Union aims to create *"an integrated continent-wide energy system where energy flows freely across borders, based on competition and the best possible use of resources, and with effective regulation of energy markets at the EU level where necessary"*.[173]

As explained in Chapter 3, the EU will need to attract new sources of natural gas supplies to compensate for declining domestic production, at least until the 2030s. The imported gas could be a mix of global LNG and pipeline gas, as well as domestic renewable gases and different forms of hydrogen. In any event, irrespective of whether the investor is a third country or EU investor, there will also be a need to close the energy infrastructure gap which is concentrated around South Eastern Europe. The price of gas and the overall project costs are critical. As stated by the former Vice-Chair of the ACER Board of Regulators and

---

[172]Eurogas views on the modification of the Gas Directive, January 2018, available at: http://www.eurogas.org/uploads/media/18PP002_-_Eurogas_views_on_the_modification_of_the_Gas_Directive.pdf.

[173]Energy Union Package Communication from the Commission to the European Parliament, the Council, the European Economic and Social Committee, the Committee of the Regions and the European Investment Bank a Framework Strategy for a Resilient Energy Union with a Forward-Looking Climate Change Policy, Brussels, 25.2.2015, COM (2015) 80 final.

Vice-President of the CEER, *"Europe needs to ensure that it keeps prices at a level low enough to be competitive. This would be impacted by how much infrastructure was built"*.[174]

There is a learning curve with every new piece of major regulation. The changes introduced by the Third Energy Package were radical, and they are just beginning to work as it takes time for market integration to progress and competition to emerge. The ACER 2016 Market Monitoring report recommended that Member States complete the transposition of the Third Package into national legislation. It also encouraged the adoption of the Network Codes, and for the Energy Union to expand cross-border links via cooperation and removing regulatory barriers.[175] A completed internal market would enable the creation of a level playing field both for external suppliers and EU companies. Fixing the internal market does not necessarily entail its over-regulation or re-regulation. Making the investment decision making, project development, construction and operational stages smooth and efficient via guidelines, streamlining the permitting processes for cross-border projects, enhancing transparency in the PCI selection process and even engaging the Commission from an early stage in the negotiations to prevent bottlenecks down the line, would facilitate implementation of the Energy Union and Security of Supply Regulations objectives.[176] The lessons learned from market creation for electricity and gas sectors should be implemented in the adoption of measures for renewable and other low carbon gases. An important pathway to achieve this is effective implementation of sector coupling which will be discussed next before once again, before turning to the

---

[174]News Article, Natural Gas World, Solidarity in the EU—The Regulators' View, 21.05.2015, available at: http://www.naturalgaseurope.com/walter-boltz-acer-eu-gas-solidarity-gie-conference-dublin-23706?utm_source=Natural+Gas+Europe+Newsletter&utm_campaign=c92b14586d-RSS_EMAIL_CAMPAIGN&utm_medium=email&utm_term=0_c95c702d4c-c92b14586d-307768681.

[175]ACER/CEER, Annual Report on the Results of Monitoring the Internal Electricity and Gas Markets in 2016, Gas Wholesale Markets Volume, October 2017, available at: https://acer.europa.eu/Official_documents/Acts_of_the_Agency/Publication/ACER%20Market%20Monitoring%20Report%202016%20-%20GAS.pdf.

[176]TYNDP 2015, 33.

2020 Gas Package and conclusions of this book. The direction of these developments is unknown at the moment: considering the decision making process, could sector coupling and the 2020 Gas Package provide an opportunity for de-politicisation of the gas sector, thus enabling an investment friendly market for decarbonisation of the gas grid?

6.10.2 Sector Coupling and Power to Gas and X Infrastructure

Sector coupling is a concept primarily used to define electrification, with renewable energy sources, of end-user sectors such as heating and transport. However, this concept has evolved in recent years to also encompass supply side sectors, mainly integration of power and as sectors through power to gas technologies like green hydrogen. In principle, this can enable cost effective decarbonisation of the gas sector.

The objective of sector coupling is to decarbonise the industry with increased us of hydrogen, biomass, biomethane and CCS. These gases can also be vital for decarbonisation of heating in buildings and in the transport sector. Sector coupling also includes the use of synthetic fuels which can be used in long-distance shipping and aviation. Another potential benefit of sector coupling is enhanced flexibility and the ability of renewable and decarbonised gases to balance fluctuations in renewable production.

Power to X refers to production of other energy vectors such as heat and liquids (this term is also used to refer to power to gas). The integration means that surplus electricity could be used to produce these sources and existing gas infrastructure can be used to transport them instead of investing in expanding the electricity transmission network (which is a very mineral intensive process). Existing gas storage can also be put in use to provide short-term flexibility. Gas fired power plants or fuel cells can provide backup capacity. This concept is supported by the Commission; however, some major technology and market barriers remain. One reason for this is that fossil fuels are still heavily subsidised and carbon pricing is not effective enough, hence there remains insufficient competition between decarbonised gases and natural gas.

The technologies need to evolve further to bring the costs down and scale up (otherwise the technology itself is mature) and existing infrastructure may need refurbishment, and new infrastructure for their generation will need to be developed. Furthermore, not every region in the EU has excess electricity. This concept is most suitable for instance in Denmark and other coastal countries with ample wind potential. Therefore, sector coupling will also have a hand in hydrogen production from natural gas and CCS.

There are also some barriers related to the access tariffs between electricity and gas networks. As such, they need to be harmonised in a way that enables a fair distribution of the associated costs between energy producers, users and energy storage service providers to incentivise the availability of balancing services and the deployment of renewable energy. Currently, there is no integrated planning in gas and electricity infrastructure. Another issue that is not yet very clear is that, taking into account the current structuring of TSOs, how will the ownership and operation of these facilities be managed among electricity and gas stakeholders? Does this mean an end to the unbundling model and independence of TSOs? The 2020 Gas Package (some suggest that it should be called the 'sector integration' or 'decarbonisation package') comes at the right time; and if the future role of hydrogen is addressed, and cost-reflective energy price signals can be developed, liquidity and markets can be created. However, political and regulatory interventions should be avoided and regulation should focus on incentivising investments in these technologies. What then are the required qualities of the relevant future regulation?

6.11 Future of Gas: Will the 2020 Gas Package Decarbonise or Subsidise Gas?

The industry and many stakeholders are calling on the Commission to develop a consistent and clear taxonomy of renewable gases and other gases that is reflective of their different lifecycle emissions (e.g. a CCS plant can be very energy and mineral intensive to develop and run). Modelling of the different potentials of these gases is also required, and a forward-looking analysis of their integration at scale to the energy mix moving towards 2050. Further, an assessment of the infrastructure needs is necessary and this calls for a revision of the TEN-E regulations cost benefit analysis methodology in particular. It also needs to reflect the importance of the DSO system as gas grids become more local (particularly with increased biomass and biomethane—depending on availability of feedstock). Finally, there is a demand to make research and innovation funds more available and accessible for competitive carbon neutral industrial solutions. This raises the following question: What is the stage of development of this yet-to-be named regulation?

Currently, ACER is carrying out a public consultation process (during the holiday period yet again within a short timeline) on the possible direction of future legislative action. At the EU level, scenario development for TYNDP are developed by NRAs. ACER is proposing to have approval power or be given the power to develop guidelines on these in line with the National Energy and Climate Plans so that competition could be introduced for decarbonised gases and multiple energy providers can also come forward, as the TSOs have a natural monopoly element and this may result in a barrier to the integration of renewable gases or other not-network based solutions. In light of criticism regarding the cost benefit methodology, ACER is seeking to develop the methodology and be assigned powers to develop guidelines so that it can reflect decarbonisation and give consideration to the development of a pure hydrogen network. ACER notes that while this may seem premature, some principles such as third-party access can already be developed to provide predictability for investments (currently hydrogen is piped via a single network to the industry user but this may change with further uptake of renewable gases).

Alongside this, it is important to note that for TYNDP 2020, ENTSOG is working with ENTSO-E for an interlinked approach to the assessment of the gas and electricity infrastructure, and they have already initiated the scenario development process in 2018.[177]

---

[177]ENTSOs joint, TYNDP 2018 Scenario Report Main report—Draft edition, available at: https://www.entsog.eu/sites/default/files/files-old-website/publications/TYNDP/2017/entsos_tyndp_2018_Scenario_Report_draft_edition.pdf.

The storylines proposed for TYNDP 2020 have been developed and consulted with stakeholders in view of publishing the final scenario report later in 2019. This should also help facilitate synergies to create hybrid energy carries as a necessary condition to achieve the climate and energy targets for 2050 in a competitive and secure way.

Another point raised by ACER is that of transparency on emissions and that LNG, natural gas, hydrogen and biomass producers should all make their emissions under a standardised methodology through the European Methane Emissions Observatory. This could enable a calculation of actual emissions and help improve calculation of the costs of different technologies.

The 2009 Gas Directive does not apply to decarbonised gases: it only covers natural gas. Saying that, ACER is recommending the possible extension of these rules with some carve-out options, and that PCIs should be significantly extended for power to X installations. It is very important here to make sure that, once again, commercial decisions are not taken by political bodies who may lack the required knowledge. Future legislation should be future proof, and technology neutral and let the market and investors decide which technology could bring the greatest value for the end-users at the lowest cost. Otherwise, the EU would continue to subside decarbonisation instead of incentivising it with the view of market creation.

6.12 Chapter Conclusions

This Chapter has revealed that since the late 1990s, the EU energy policy has undergone substantial modification, with the Second and Third energy packages bringing some radical changes to how natural gas markets operate in the EU. This has resulted in increased control of the market and regular intervention by the Commission in market operations. This Chapter has underscored the importance of regulatory certainty and clarity, and referred to the areas where these elements diminish, such as when exemption decision are taken under Article 36 of the Third Energy Package. This provision, together with the information exchange regulations and 2019 Gas Directive Amendment indicates that natural gas import projects will continue to be motivated by political rationale rather than commercial interests. At this point in time, the majority of most cross-border gas infrastructure projects are heavily subsidised from the EU budget. This makes attracting commercial investment into large-scale energy supply infrastructure in the EU difficult. This may be detrimental for energy security in the short term as indigenous production is falling rapidly in the EU.

This chapter also included a discussion on the external aspect of EU energy policy. Supply security and transit interruptions (between 206 and 2014) had a significant impact on the development of EU policy in this sphere. It is difficult to divorce regional and national aspects of energy from the external dimension. The EU's current import dependency requires an energy policy built on close cooperation with international partners and based on mutually accepted principles. This is why EU energy policy is part of its foreign policy. The Energy Union as a whole provides an opportunity for the identification and promotion of specific international cooperation frameworks to strengthen relations with producer and transit States outside the EU (which can in the near future facilitate cross-border hydrogen transport) but this process is far from perfect. There is also substantial potential for the EU to take the lead in proving the concept of gas sector decarbonisation and strengthen its role in global energy markets.

Last but not least, this chapter covered the final remarks on the expected 2020 Gas Package and the future of gas in the EU, which has potential as sector coupling is supported by the EU institutions. The key to its success is to adhere to the market principles which were introduced for natural gas in the first place but perhaps not implemented at full-scale, largely as a result of the shift in decision making on infrastructure projects. There are so many unknowns regarding what the future energy mix may look like towards 2050, and more could be planned in advance if an integrated and inclusive approach is employed in adoption of different policy frameworks and financial incentives. The next chapter will conclude this book, reiterating that beyond 2050, and perhaps even before that, natural gas will be a high carbon energy source in the EU, but action must be taken now.

7

Conclusion

7.1 The Future of Gas in a Changing World

The discussion in this book was not on whether gas investments are counterproductive or not, as it argues that gas is likely to remain in the energy mix until viable market-ready solutions for heating and transport exist at scale. Of relevance, in the EU indigenous gas production continues to fall in all major gas producing Member States with the largest gas field in the Netherlands (Groningen) producing 11% less gas in 2018 compared to the previous year. In total, in the first quarter in 2019, gas production fell by 31 bcm.[1] Demand, in contrast, has remained at roughly the same level for the last five years. As result, there was an increase in LNG imports and a new major pipeline allowing access to resources in the Caspian region will come online in 2020. The decline in production and a number of repeated interruptions to flows of natural gas from the EU's main supplier, Russia, instigated a

---

[1]Quarterly Report Energy on European Gas Markets Market Observatory for Energy DG Energy, Volume 12 (issue 1, first quarter of 2019), available at: https://ec.europa.eu/energy/sites/ener/files/quarterly_report_on_european_gas_markets_q1_2019_final.pdf.

© The Author(s) 2020
G. Mete, *Energy Transitions and the Future of Gas in the EU*, Energy, Climate and the Environment, https://doi.org/10.1007/978-3-030-32614-2_7

diversification policy which intensified in 2014 and remains relevant today. In addition to diversification, the EU, as exemplified in this book, implemented a number of measures to increase security of supply via investments in infrastructure links that enables reverse flows so that in case of emergencies, supplies can be diverted within the EU. Furthermore, new natural gas import facilities in regions with access to only one supply route and in energy islands have been identified and commissioned with the support of EU funds. What this book has shown is that these strategic projects were not commercially viable. In other words, the rate of return of investment is either too low or resulting in losses. Industry associations, most visibly the European Federation of Energy Traders (EFET) have stated on a number of occasions that they do not *"support projects that are not backed by a positive cost-benefit analysis,* as such projects would be counterproductive for European gas *market integration".*[2]

This book provides a detailed account of the financial, infrastructure-related and regulatory elements of gas infrastructure projects. It demonstrates that within the EU internal market; pipelines can be constructed as a merchant pipeline, or the costs are socialised (which in effect would increase transmission tariffs).[3] While normal investments, where the capital costs are repaid by transmission tariffs, are often undertaken by the TSOs, the merchant investments, which are instead used for cross-border pipelines (within the EU or connecting with a third country) are generally made by non-TSO investors.[4] Article 13 of the Gas Regulation sets an obligation on NRAs to ensure that the tariffs provide appropriate return on investments and that there are sufficient incentives to construct new gas infrastructure. However, this book has

---

[2]The European Federation of Energy Traders (EFET), Greece–Italy Incremental Capacity Proposal, 18.12.2017, available at: http://www.efet.org/Files/Documents/Downloads/EFET%20 comments%20incremental%20capacity%20Greece%20-%20Italy.pdf.

[3]Jonathan Gaventa, Manon Dufour, Luca Bergamaschi, More Security, Lower Cost a Smarter approach to Gas Infrastructure in Europe, Energy Union Insight Series, E3G, March 2016, available at: http://www.energyunionchoice.eu/wpcontent/uploads/2017/08/E3G_More_security_ lower_cost__Gas_infrastructure_in_Europe-1.pdf.

[4]Kim Talus, Decades of EU Energy Policy: Towards Politically Driven Markets, *The Journal of World Energy Law & Business*, Volume 10, Issue 5, 1 October 2017, pp. 380–388.

shown that all of the PCI gas pipeline projects have required financial backing from European institutions.

The gas infrastructure case studies provided here serve to emphasise the importance of predictability of law and policy and open markets for enabling private investment in the gas value chain under the principles of project finance. Despite this, increasing political interference with the market results in the potential to damage the occurrence of commercial investment decisions in the EU, particularly foreign investment in the energy infrastructure sector. This book suggests that where there is a functioning market, the risks should be borne by the private sector. They can then assess whether to invest in LNG, storage and pipeline infrastructure, or diversify their portfolio in renewable energies. Given the urgency of addressing climate change in a very strict timeframe, EU policy must act to not only support security and diversity of supply concerns (which are important) but also to incentivise gas infrastructure projects for renewable gases, hydrogen and CCS/CCUS.

One of the main debates elaborated upon in this book is that regulation, with regard to the financing challenges that project sponsors face, should take the commerciality of projects and the depoliticalisation of decision-making processes for energy investments in the EU into account. Allowing the market to take the risk would remove the need to subsidise strategic projects, leaving room in the EU budget for the more pressing issues faced by the regional economic bloc as a whole and invest the funds into a green economy. EU investment policy makers may like to reconsider their priorities in investment, in order to achieve more efficient and flexible markets, and decide between market-based approaches or a market that is regulated; the current situation which leaves the project sponsors in limbo may not be the best solution for targeted diversification of supply.

This book concludes that gas infrastructure has the potential to be a vital component of the EU's energy security and its commitment to the climate change towards 2050. It is stated multiple times here that gas sector decarbonisation is inevitable, and various types of renewable and decarbonised gases can play an important role. The existing pipeline network can be used for future energy systems without the need for any major modifications. By providing timely real-world case studies, this book provides examples, for instance, to show that it is possible to inject a mixture of conventional and unconventional gas into the system, or hydrogen generated with electricity from renewable energies and methane from biomass. Hydrogen is already used as a gas and liquid in the petroleum industry and in manufacturing processes for producing chemicals, food-types and electronics. The EU regulates the required quality of gas produced from hydrogen and biogas in order to be allowed access to the natural gas network. Furthermore, it has been noted that gas has substantial potential to decarbonise the transport sector, in particular compressed natural gas (CNG) for heavy duty/long distance transport and LNG as marine fuel. Examples already exist of the use of hydrogen as a fuel in the public transportation sector, including from Germany and the United Kingdom. In the aviation sector, attempts are being made to develop a hybrid diesel/hydrogen fuel cell engine.

In the EU, elements relevant for hydrogen technologies and biomass, beyond setting the general level of ambition for decarbonisation, include guarantees of origin for hydrogen and biomass, and regulatory treatment of renewable gases in transport. From a legal perspective, risks associated with injecting hydrogen and biomass into the existing network of gas infrastructure can be addressed via enhanced regulatory capacity, including technical expertise and the development of a sound legal framework that keeps up with innovation. There are currently only small amounts of green hydrogen and biomass production in the world. Advancement of this sector may require some form of subsidisation. This could be avoided if dynamic pricing of network and energy tariffs (i.e. electricity and gas tariff competition) and investment incentives are introduced by legislation. However, if regulation is adopted without proper cost-benefit analysis, flexibility and consultation with the stakeholders, it may not be adapted to the real-world state of affairs which is still in flux and may convey a wrong message to the public about the advantages or security of virtual currencies.

In November 2018 the Commission published 'Clean Planet for All' report, looking for carbon neutrality across the EU by 2050. This sets the future direction of travel of EU energy and climate policies, translated into legislation under the 'Clean Energy Package', which is acting to drive forward the EU's climate agenda. However, even if this is attained, there will likely remain some emissions in industry and agriculture that have to be compensated by natural processes or carbon removal. In the last two Paris-compatible scenarios set out in the report, energy demand is reduced by 25–30% while demand for gas energy vectors is limited. Therefore, the role of gas in the energy sector will inevitably be reduced in the future, however gas and hydrogen are likely to remain in the industry sector, and zero-carbon gases may be used in long-distance transport by truck and ships. Between now and 2050, gas will continue to provide flexibility to the energy system, and increased sector integration between electricity and gas could help significantly with the storage of large quantities of imported renewables.

Another conclusion from this book is the importance to develop an integrated vision for gas in the future at the EU level; Member States should develop their plans in alignment with the EU objectives. However, this strategy should be adapted depending on the level of technological and economical progresses. In 2018, for the first time ENTSOE and ENTSOG worked together to produce a joint report on European scenarios for gas and electricity infrastructure to facilitate connecting and integrating the two key sectors and develop new clean energy scenarios. The support provided to sector coupling at the policy level by the EU institutions are positive developments in this regard. This said, the development of zero-carbon gas (biogas, hydrogen, CCS, synthetic gas) should be carefully developed to avoid indirect impacts on the environment (such as CO_2 leakage).

This book introduced different types of decarbonisation pathways. None of them offer a silver bullet as of yet. Biogas and methane are carbon free however they do not offer sufficient solution in terms of volumes, as feedstock, waste and agriculture sources are limited. They also face the problem of not being able to be injected into the TSO grid; instead, they offer local solutions via the DSO grid because biogases have to be purified and upgraded which is a very costly process. CCS presents us with a unique opportunity to maintain the competitiveness of European industries and create new jobs and economic benefits but again it is not a stand-alone silver bullet. It is a climate mitigation solution but not an energy solution. When we look at hydrogen, from an infrastructure point of view, while the industry can switch to green and low carbon gases, consumers switching to hydrogen may not be easy. The low carbon transition journey means there will be a need to find new markets. In the future, every supplier should have access to each producer unless the system is incompatible. Sustainability is the driver but it has to be viewed within the requirement of affordability and security of supply.

Therefore, this book also acknowledges that the EU cannot produce zero carbon energy at a low cost. The Sustainable Finance Package will be influential in attracting investment for clean and green energy and there will be more green jobs generated from developing and deploying these technologies. However, hydrogen, biomass and CCUS may require even higher subsidies to develop, but at least these projects are aligned with the EU's energy transitions agenda. All large energy infrastructure projects require sizable capital investments and thus need to ensure long-term revenue guarantee. Regulation going forward, therefore should have the objective to ensure market facilitation, ensuring the complete implementation of the regulatory model and provide stability to both market participants and national governments and regulatory intervention should only be allowed if the benefits of the intervention can be shown to outweigh the costs, which has so far not been the case in the natural gas sector.

Further, this book has discussed various short-, medium- and long-term projections and the future role of gas over a 50-year horizon. It concludes that the future energy system will be made out of hybrid energy carriers, where gas and electricity production, transport and storage will be more connected and linked. This transition will continue to take place gradually. The first transition that took place in the EU since 2015 is from oil and coal to gas switching and, after 2020 it will be natural gas to green gas (blue and green hydrogen), and next, after 2030 it will be CCS (although such milestones are not set in stone, dependent as they are on technological development, appropriate regulatory, financial and political initiatives, etc.). In the short term, considering the lack of immediate alternatives and continued use of oil and coal (with significant variations in volumes within the EU Member States), the role of natural gas as a lower carbon emitting source to bridge the gap between high and low carbon sources in the on-going energy transition should not be undermined. However, if the 2050 targets to be achieved, limited public resources must now be directed towards new infrastructure that will be needed for a more diversified mix of renewable and low carbon gases. And the problems in addressing rising temperatures are formidable. At the local level we are not feeling climate change so we are not changing our lives yet. Different governments and departments in the public space also do not cooperate. It is hoped that this book will contribute to raising awareness on this issue, which is timely given warnings of the dangers of climate change.

Finally, it is important to underline that we are living in a dependent world that is interlinked. The EU is not an island. The EU has energy relations with its oil and gas producer neighbours, for instance in the Mediterranean. There will be a need to establish new relations with the major suppliers. How the EU can decarbonise with other countries, and bring them onboard to commit to CO_2 reduction should also be included in the future policy of the EU, and be part of its forward-looking strategy.

7.2   Recommendations

Numerous specific recommendations have been made in this book; this section focuses on the broader recommendations with implications for the future of the gas sector in the EU as it addresses climate change in the ongoing energy transition.

1. The EU approach to natural gas infrastructure should not only focus on security of supply and diversification concerns, but also address climate change and affordability as decarbonisation will not come cheap nor will it happen overnight.
2. In this respect, this book recommends that the EU should implement a two-tier system—one addressing short to medium term challenges and one for the future of its energy architecture. Further interim investments in natural gas infrastructure are still needed today to complete the internal market. EU action taken on this now

is currently sub-optimal. If decisions are left to the market the outcomes would be more efficient, and it is recommended here that the EU should use these efficiencies, rather than continuing to use public money which could otherwise be capitalised by market participants. Greater public resources could then be made available to promote renewable gases and facilitate climate friendly and future proof energy infrastructure.

3. The time to get the 2020 Gas Package right is now. Any later could be catastrophic. It took over 20 years to create an internal market for gas and electricity, and it is still evolving. The IPCC recently reported that the world has only 12 years to cut fossil fuels to avoid devastating global warming. Hence, the time frame for decarbonization of the EU economy, enabling a future role for renewable and other clean gases may be stricter than it is currently foreseen.

Index

© The Editor(s) (if applicable) and The Author(s), under exclusive license to Springer
Nature Switzerland AG 2020
G. Mete, *Energy Transitions and the Future of Gas in the EU*, Energy, Climate
and the Environment, https://doi.org/10.1007/978-3-030-32614-2

Printed by Printforce, the Netherlands